Soldier of Fortune

Soldier of Fortune

*Adventuring in
Latin America and Mexico with
Emil Lewis Holmdahl*

⚜

by
Douglas V. Meed

Halcyon Press Ltd

Published by Halcyon Press, Ltd. www.halcyon-press.com

© Copyright 2003 by Douglas V. Meed

All Rights Reserved.

No part of this book may be used or reproduced in any manner whatsoever without prior written permission from Halcyon Press Ltd. No part of this book may be stored in a retrieval system or transmitted in any form or by any means including electrostatic, electronic, magnetic, mechanical, photocopying, recording, or otherwise without the prior written permission of Halcyon Press Ltd Houston, Texas.

For information, write:

Halcyon Press, Ltd.
6065 Hillcroft Suite 525
Houston, TX 77081
info@halcyon-press.com

First Edition

Library of Congress Cataloging-in-Publication Data

Meed, Douglas V.
  Soldier of fortune : adventuring in Latin America and Mexico with Emil Lewis Holmdahl / Douglas V. Meed.
    p. cm.
Includes bibliographical references (p.) and index.
  ISBN 1-931823-05-7 (alk. paper)
  1. Holmdahl, Emil Lewis. 2. Soldiers—United States—Biography. 3. Mercenary troops—Latin America—Biography. 4. Mercenary troops—Mexico—Biography. I. Title.
  U53.H644 M44 2002
  355'.0092—dc21

2002011345

For Terry and Joan McCollister
El Paso, Texas.

* * * * *

This book could not have been written without the help of the following: Thanks to Gordon and Rose Holmdahl; Robert Stonedale, Palacios, Texas, and Edrena Jones, El Paso, Texas; my family - Jeannine, Mike, Sonia, Alex, and Geoff. Also thanks to the librarians at the Bancroft Library, University of California at Berkeley; University of Texas at El Paso; the Center for American History and the Benson Latin American Collection at the University of Texas at Austin; and the El Paso Public Library for their assistance.

Contents

List of Illustrations vii
Preface xi

CHAPTER   PAGE

1. Benevolent Assimilation 1
2. Fighting the Moros 19
3. Earthquake in Old 'Frisco 29
4. The Banana Men 35
5. Revolution in Mexico 45
6. Into Yaqui Country 61
7. The Attila of the South 71
8. The Orozco Revolution 83
9. Riding with Pancho Villa 99
10. Trial and Redemption 115
11. Young "Blood and Guts" 133
12. With the A.E.F. 155
13. Drifting 169
14. La Cabeza de Pancho Villa 177
15. Keep Coming On 189

Notes 198
Bibliography 212
Index 220

❧ Illustrations ❦

Emil Holmdahl, 15 years old, 1898 x
Map 1: The Philippines xiv
Emil Holmdahl during the Philippine Insurrection 2
Emil Holmdahl and comrades, c. 1901 18
Emil Holmdahl in his dress uniform 22
San Francisco after the 1906 earthquake 28
Map 2: Central America 34
"Tex" O'Reilly 40
Map 3: Northwestern Mexico in the 1910s 44
Emil Holmdahl in Mexico, c. 1911 48
Yaquis in Sonora, 1913 60
General Benjamin Viljoen 62
Giuseppe Garibaldi 64
Yaqui camp followers, 1913-1914 66
Emil Holmdahl and T.J. Beaudit, Feb. 1912 69
Emil Holmdahl and dog on horseback, 1910s 70
Emiliano Zapata's Pistol 81
Pascual Orozco's Irregulars 82
"Colorados" or "Red Flaggers" 85
Sam Dreben 87
Tracy Richardson 88
Pancho Villa and his bandits, 1911. 98
Villa's Cavalry entering Juárez, 1911. 103
Obregón, Villa, and Pershing. 114
Map 4: General Route of the Punitive Expedition. 132
The U.S. 7th Cavalry departing Ft. Bliss. 134
General John "Black Jack" Pershing. 136

U.S. Cavalry in Chihuahua during the punitive expedition 137
Lt. George S. Patton. 140
One of Patton's Dodge Touring Cars. 141
Sketch of Rubio Ranch. 142
The Palace Hotel. 144
Open ground at Carrizal. 151
Emil Holmdahl during World War I 154
Emil Holmdahl in Mexico, 1920s. 168
The death mask of Pancho Villa 176
Emil Holmdahl in his later years 188
Emil Holmdahl in 1962. 195

Emil Holmdahl was fifteen years old when he enlisted in the U.S. Army during the Spanish-American War in 1898. — Holmdahl Papers

❧ Preface ❦

*I drew these tides of men into my hands
and wrote my will across the sky in stars.*

— T.E. Lawrence

Emil Lewis Holmdahl was the last of that fabulous breed of soldiers-of-fortune who swashbuckled their way through the wars and revolutions at the beginning of the 20th century before the romance of soldiering died in the muddy, blood-soaked trenches of World War I.

Holmdahl was born in Fort Dodge, Iowa, to a farming family of Swedish immigrants on August 26, 1883.[1] At an early age he thrilled to Rudyard Kipling's stories of exotic battles in the Far East and to the "dime novel" adventures of cavalrymen and lawmen in the American West. But unlike most boys, he turned his childhood dreams of martial glory into an exciting, if dangerous, reality.[2]

Escaping his rustic origins, from age fifteen to eighty, he swaggered his way across a score of battlefields. He soldiered in the mountains and jungles of Asia, through the swamps and crumbling ancient cities of Latin America, in the ferocious battles of the Mexican Revolution, and in the hell of World War I trenches. He fought Filipino insurgents under the Stars and Stripes, overturned Central American dictators, battled alongside Pancho Villa, then fought against Villa, and probably was the man who cut off Villa's head.

Holmdahl was condemned to a U.S. federal prison for gunrunning until paroled to serve General John J. Pershing during the 1916 punitive expedition into Mexico. There, he guided a green Second Lieutenant George Patton across the Chihuahuan desert to his first bloody gunfight. After a pardon granted by President Woodrow Wilson, Holmdahl fought as a commissioned officer alongside American and British soldiers in France in World War I. There he helped smash the last German offensive in 1918. As an aged senior citizen, he was investigated by the Secret Service for smuggling gold bars out of Mexico.

Although he had the hard-edged roughness of the professional soldier and hardly any formal education, Holmdahl held the respect and affection of generals and political leaders on both sides of the U.S.-Mexican border. In some of his more introspective writings he showed a touch of the poet. Surprisingly, for all his "rake-hell" adventures, this tough and complex man managed to live to a ripe old age.

It was through a few very fleeting mentions of his name in El Paso newspapers during the years of the Mexican revolution that I first came upon the Holmdahl story. Intrigued, I searched through many histories, memoirs, and newspaper accounts of the revolution and found only a few tantalizing tidbits of information. A thorough mining of El Paso newspapers revealed little except stories of his 1915 trial for gunrunning.

His army records from his first enlistment in 1898 through World War I were destroyed by a fire in the St. Louis Federal Depository. Records of his service as a civilian scout and spy for General Pershing have disappeared—possibly destroyed on that general's orders. Remaining are the personal letters of 2nd Lt. Patton, General Pershing, and verbal reminiscences of those who knew him. His reports to the War Department are preserved in the National Archives.

While he gave a few contemporary interviews during the Mexican Revolution to Chicago and California newspapers, for the most part he was a secretive, furtive figure. This is not surprising since he was often involved in gunrunning, spying, revolutionary plots, and secret missions along the Mexican border. Like many of

Preface

the freebooters who fought in the Mexican revolution, there are numerous legends about him sung in the *corridos*, the folk songs of Mexican peasants. Surprisingly, many of them are true.

Fortunately, after Holmdahl's death in 1963, his nephew, Gordon R. Holmdahl, bundled up his uncle's vast array of newspaper stories, United States and Mexican government documents, and diary jottings. These, in addition to other records he held, cast much light on his uncle's sometimes flamboyant and often shadowy lifetime adventures. Those records and a large number of photographs were donated to the Bancroft Library at the University of California at Berkeley. My wife and I flew to Berkeley and were the first, and perhaps the only, persons to examine the material. At his home in California, Gordon Holmdahl provided other insightful material and a treasure trove of photographs.

This is the story of that fabled man, Emil Lewis Holmdahl. He was the last of the great soldiers-of-fortune who roamed the world fighting under different flags for money, adventure, sometimes for principle, but mostly just for the hell of it.

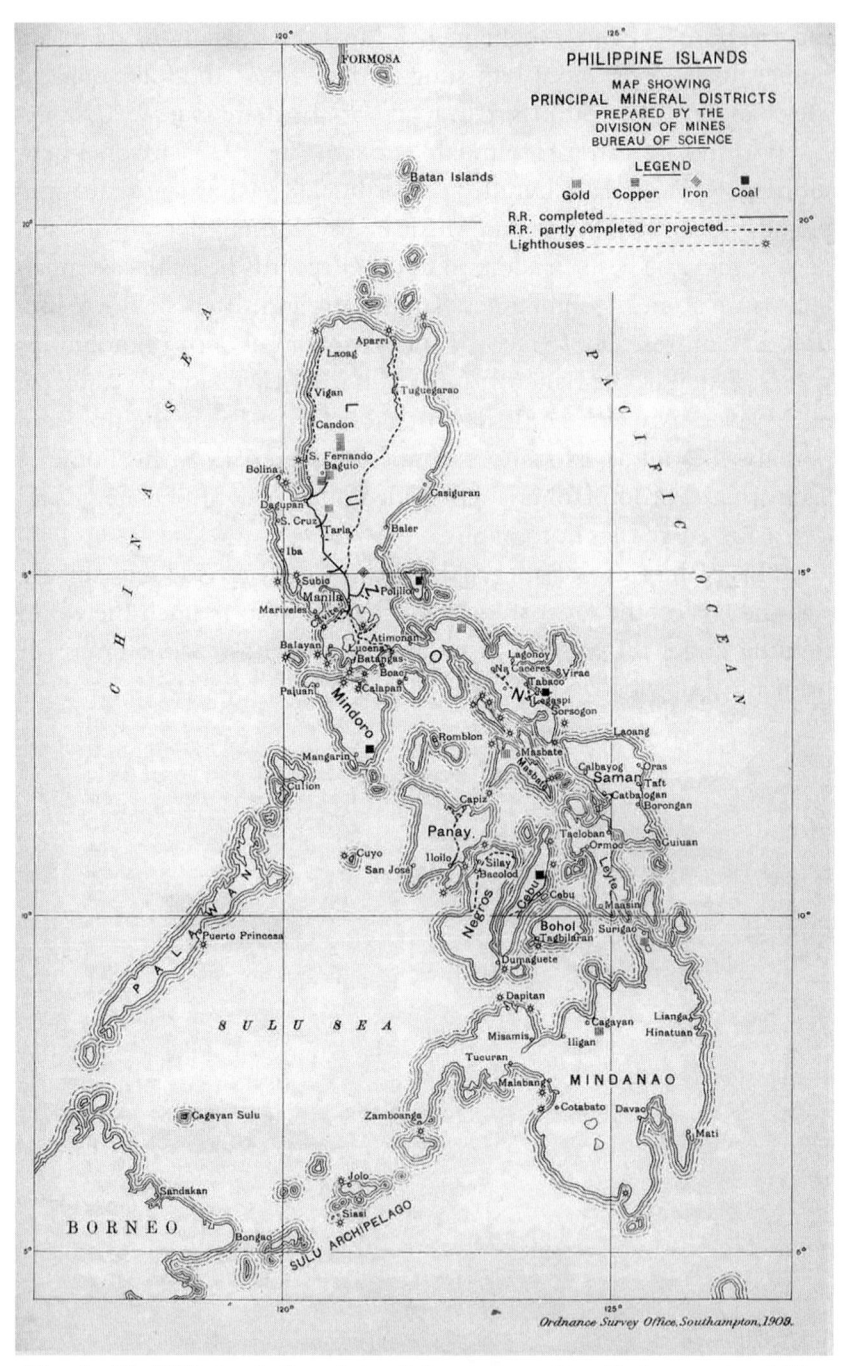

Map 1: *The Philippines in the early twentieth century.* — Provided courtesy of The General Libraries, The University of Texas at Austin

1
Benevolent Assimilation

"Damn, Damn, Damn, the Filipinos,
Underneath the starry flag,
Civilize 'em with a Krag."
— U.S. soldiers' song

At the outbreak of the Spanish-American War, President McKinley on April 23, 1898, called for 125,000 volunteers to beef up the anemic U.S. Army of only 28,000 regulars. Emil Holmdahl's older brother, Monty, was one of the first to join up. With a yen for adventure and not wanting to be left behind, young Emil also trooped down to the recruiting station.

Although tall for his age, the fifteen-year-old was told by a grizzled old sergeant to go back to the farm and do his chores for a few more years. Even at this early age, young Holmdahl had a touch of the confidence man. Undaunted, he took his small savings and went to a different recruiting station, where he hired a man to pose as his father and testify that he was of age.

The ruse worked and the slender, apple-cheeked farm boy was sworn in as a rifleman with the 51st Volunteer Iowa Infantry Regiment.[1] After a brief training period, the Iowa boys lustily singing, *"Remember the Maine, to hell with Spain,"* were put on board a train en route to San Francisco. There on November 3, 1898, they boarded the crowded troop transport *Pennsylvania* for a month-long passage to the Philippine Islands. They arrived in Manila Harbor on

December 7. By then the Spanish had surrendered and the "Splendid Little War," so named by Secretary of State John Hay, had ended.

If Emil was disappointed that he had missed action against the Spanish army, he had not long to wait for his baptism of fire. For there was smoldering resentment between the ragged, irregular forces of the Filipino insurgent leader Emilio Aguinaldo, who had been fighting against the Spanish since 1896, and the American interlopers.

Most Filipinos maintained they had virtually won their independence before the Americans arrived, and they rejected the idea that they needed American guidance. They claimed, "When the American troops reached the Islands in 1898, there was no anarchy and the Filipinos were governing themselves."[2] The idea that Americans had single-handedly liberated the Philippines from Spain was a constant irritant to the Filipinos.

Emil Holmdahl during the Philippine Insurrection.
— Holmdahl Papers

The Filipinos expected independence after the Spanish were beaten, but found themselves merely trading colonial masters, as President McKinley instituted a policy of "benevolent assimilation." This term meant that the United States would control the islands. No matter how benevolent the Americans were, Aguinaldo and his men would have none of it.

The McKinley policy was not without opposition in the United States. Republican Speaker of the House, Thomas B. Reed of

Maine, opposed the war with Spain and the acquisition of the Philippines. When Vermont's Senator Redfield Proctor read a report on the Senate floor favoring the war, Reed observed that the senator owned large marble quarries. "Proctor's position might have been expected," he said. "A war will make a large market for gravestones."[3]

Not only did opponents decry American imperialism, there were financial concerns. After Spain surrendered and the United States seized the Philippines, Reed cynically observed, "We have purchased 30 million Malays at 50 cents a head . . . unpicked . . . and nobody knows what it will cost to pick them."[4] Acquiring the Philippines would prove very expensive, both in money and in lives.

The growing disputes between the Americans and the Filipino nationalist forces under Aguinaldo, however, were unknown to Emil and the rest of the Iowa farm boys of the 51st. After a month at sea they were yearning for the feel of solid earth beneath their feet. From the decks of the now smelly *Pennsylvania*, they could see the lights of the grog shops and bordellos of Manila and they were hot for a little Asian debauchery. They would never get it.

After days of staring at the shoreline while the Army pondered what to do with them, orders finally came through on December 24. The *Pennsylvania* was instructed to proceed to Iloilo on the island of Panay, where the troops would be landed and the Iowa boys would take possession of the port to "prevent lawlessness."[5]

Panay, with a population of almost 800,000, was part of the untamed and remote group of islands known as the Visayas. On December 26, escorted by the cruiser *Baltimore*, the *Pennsylvania* and two other troopships sailed out of Manila Bay into the South China Sea, passed between the islands of Mindoro and Palawan into the Sulu Sea, and finally dropped anchor off Iloilo.

As the ships swayed at their anchor cables, the Iowans panted on deck under the boiling tropical sun while the high command dithered over whether or not to invade the island, then under the control of Filipino insurgents. On January 3, orders came through to attack and capture the city. That night Emil and his comrades sharpened their long bayonets and cleaned their old .45-70 Springfields (which when fired sent up a dense column of white-gray smoke) with the nervous energy of green recruits about to undergo a baptism of hostile fire.

The next morning the 51st, with units of the 18th U.S. Infantry Regiment and Battery G, 6th U.S. Field Artillery, clambered down the sides of the transport into longboats and were rowed ashore by strong-armed sailors. As they landed, they were greeted by insurgents barricaded along the docks with Mauser rifles at the ready. The two forces glared ferociously at one another for a few minutes until the American colonel in command decided against landing more troops "under such conditions of hostility."[6]

In a scene more *opera bouffe* than military, the disgusted troops were ordered to climb back into the boats, and the equally disgusted bluejackets rowed them back to the transports. Dithering between the Panay Expeditionary Force and headquarters in Manila went on for twenty more days, while the troops rocked uneasily in the anchored transports in the stifling heat.

After almost three months aboard ship, the Iowans were so enervated, filthy, sick, and dispirited that the *Pennsylvania* was ordered back to Manila on February 11, 1899. Returning to that port in mid-February, they stumbled down the gangplanks and found themselves right in the middle of a shooting war. The Iloilo farce over, they were to engage in vicious and grueling fighting on southern Luzon Island as the U.S. army began its breakout from its Manila bastion.

Of young Holmdahl's personal adventures in the Philippines little is known, but one can best understand the conditions that molded the man by following the trail of his regiment. On the outskirts of Manila, Yanks and insurgent soldiers had been exchanging insults and sometimes blows, until on February 4, 1899, a Nebraska volunteer fired on a Filipino patrol, and the two forces went to war.

It would be a dirty and frustrating guerrilla war fought in steaming jungles rife with malaria and other tropical diseases. The troops were tortured by hordes of stinging and biting insects, poisonous snakes, and an enemy that posed as friendly during the day only to spread terror and death as soon as the tropical sun dipped below the horizon. It would only be surpassed more than sixty-five years later in the hell of the Vietnam jungles.

Into this cauldron marched the 51st Iowa farm boys, including fifteen-year-old Emil. He was to celebrate his adolescence by wading through rice paddies, tramping along jungle trails, and soaking in tropical downpours, all on a diet of rice and rotten fish when supplies gave out. In the realm of nauseating food, Army-issue canned salmon was king. Called "goldfish" by the troops the stuff was so full of oily gunk that when it dried out and a lighted match was tossed onto it, the can burst into flame. By comparison World War II's K-rations so despised by G.I.s seemed a gourmet's delight.[7]

A typical order from a brigade headquarters ran as follows:

> *Men will carry guns with straps and bayonets, belt, haversack, mess kit, canteen filled with water or coffee. One day's field ration, 100 rounds of ammunition, poncho hung in belt. They will wear brown canvas uniform including blouse without blue shirt. Those not provided with blouses will wear blue shirts.*
>
> *Two days additional field rations, 200 rounds additional ammunition, one blanket for each two men and necessary cooking utensils, tools, etc., will be transported in wagon and pack train. Reveille will be at 3:00 a.m.; breakfast at 4:00 a.m.; and troops will be in assigned positions ready to start by 5:00 a.m. when each regimental commander will send a messenger to brigade commander to that effect. There will be no bugle calls, loud commands or shouting.*

And as a very necessary precaution, the orders commanded:

> *Officers and non-commissioned officers will prevent men from throwing away accoutrements, rations, water and ammunition.*[8]

In March 1899, Emil and the 51st were attached to the First Division under the command of Major General H.W. Lawton, who was the "George Patton" of what became known as the Philippine Insurrection. Lawton, at six feet, four inches, towered over his men and must have seemed like an avenging giant to the slightly built Filipinos. He was a soldier's soldier. With iron gray hair, bristling mustaches, and a fierce gaze, he wore a white pith helmet and bright yellow slicker, and was given to striding up and down the firing line wherever the fighting was at its heaviest.

Bold and tough, he was often held back by the overly cautious General Elwell Otis, who commanded the Philippine department with all the bravado of a frightened rabbit. Balding with floppy muttonchop whiskers, Otis was a desk-borne officer who Admiral George Dewey once called "an old woman."[9]

An orphan, Lawton left school to enlist in the Union Army as a private when the Civil War broke out in 1861. Always in the thick of the fighting, by the time he was twenty-one years old he was commanding a regiment of infantry as a brevet colonel. He won the Congressional Medal of Honor for leading an attack against a Confederate fortification at Atlanta.

After the war, reduced to the rank of Second Lieutenant, he spent twenty years fighting Indian hostiles. He was the man who finally cornered the elusive Apache chief Geronimo, who, exhausted by Lawton's indefatigable pursuit, surrendered in 1886. He again distinguished himself during the campaign in Cuba in 1898 and, by then a major general, he was assigned to help break the back of the Filipino resistance. While he was often in trouble with the high command in Manila, the men in the ranks loved the fifty-six-year-old general.

After the Americans fought their way out of the environs of Manila, the brigade, which included Emil and the 51st, also comprised elements of the 4th Cavalry, the 14th U.S. Infantry, and a regiment of Idaho volunteers. After the breakout, they slugged their way fifty miles to the north toward the key railroad town of Calumpit.

The town was enclosed in a rectangle formed by the railway that ran from Manila to San Fabian on Lingayen Gulf and three rivers that arched around the city flanks. Calumpit was well fortified with trenches and loop-holed breastworks built along the river banks, while the railroad embankment was built up with parapets enabling defenders to fire in all directions. Confident, the Filipinos considered the town impregnable and their commander boasted "Calumpit will be the sepulcher of the Americans."[10]

On the morning of April 23, the Americans approached the Quingua River to the south of Calumpit and came under a fierce and deadly fire from the dug-in Filipinos. As a regiment of

Benevolent Assimilation

Nebraska volunteer infantry formed and advanced toward the enemy, their Colonel, John M. Stotsenberg, one lieutenant, and two privates were shot dead and thiry-one men wounded. Seeing the Cornhuskers in trouble, Emil and his Iowans fixed bayonets and charged forward with a shout. Firing and running, they stormed the enemy and with rifle butt and bayonet, slaughtered the insurgent forces within.

Panting, the Mid-Westerners flopped on the soggy ground, sucking air into gasping lungs. After a rest, they marched into the little suburb of Quingua where they made camp, and rations, such as they were, were distributed as well as fresh bandoliers of ammunition. "Get a good rest boys," they were told, "Tomorrow we cross another river under fire and by god, then we'll take Calumpit."

That night, as fifteen-year-old Emil rested his weary head on his pack and looked up at the stars overhead, he must have realized he was no longer an innocent farm boy. He had seen friends killed, and he had tasted blood with both bullet and cold steel. He would never be the same.

Before dawn on April 24th, the troops were awakened from their uneasy sleep, ate a quick breakfast of biscuits and hot coffee, and then fell in with extra bandoliers of ammunition looped over their shoulders. By 5:00 a.m. the artillery was in place and ready to fire. The night before, scouts had located a ford and the infantry was moved up and echeloned along its banks. At 5:30 a.m. the artillery opened up a fierce barrage on the insurgent trenches on the far side of the river.

The South Dakota regiment of volunteers dashed across a rickety bamboo bridge, while Emil and his Iowans, along with the Nebraska volunteers, charged into the ford, slipping, stumbling, and sometimes sinking into the river's muddy bottom. Gone was the boyish enthusiasm of the day before. They had seen too many bodies on both sides crumpled into bloody rags from Mauser and Springfield slugs. They now had the grim determination of veteran troops, and they flayed the insurgents with rifle fire as shrapnel burst on the enemy trenches ahead of them.

They were across the river by early light and had advanced through dense, thorny brush and thickets of bamboo. Trudging along the road toward Calumpit in late morning and early afternoon, they were continually harassed by snipers and by the retreating insurgents. Reaching the Calumpit River, which was that town's last defensive shield, the Iowans came under heavy fire from defenders on the other side of the river. The American artillery, however, blasted the enemy trenches with shrapnel so accurately that the insurgents became afraid to raise their heads above the emplacements to aim their Mausers.

Emil and the Iowans could see hands holding rifles appear over the trenches and fire without aiming. As a result, their bullets flew wildly, mostly over the heads of the Midwestern volunteers. The American infantry, in turn, dropped prone and delivered such a stream of accurate fire from their Springfields that it all but annihilated the remaining defenders. Emil and his companions crossed the river unscathed and marched into the town under an eerie light from the Calumpit mission church, which had been set on fire by the fleeing Filipinos.

After a few day's rest, on May 2, the 51st, backed up by two field pieces and a Gatling gun, was detailed as the brigade advance guard. The troops marched along the railway running northwest to San Fernando along flat land cut by steamy swamps and bayous, housing myriads of insects.

As they moved up the road paralleling the tracks, they first encountered the traps which during the Vietnam War were called punjis. They were conical pits dug into the road, in the bottom of which were planted sharpened bamboo stakes dipped in feces. A light bamboo mat covered with dirt concealed the hole, and God help the poor American lad who stepped on one of the devilish stakes. Infection from the feces covering the stake often was more dangerous than the wound caused when the sharpened bamboo pierced a boot and drove up into a foot. Many died or lost a leg through gangrene.

But if what had gone before was not bad enough, a few miles south of their objective, in the market village of San Fernando, they ran into rifle fire from a swamp near the town. To drive the enemy from their flanking position, the Iowans were ordered to charge.

Dismayed but game, the 51st waded into the foul-smelling mess, floundering through the treacherous waters, often sinking up to their armpits. Leeches sucked at their bodies; their water-filled boots caused them to stumble and fall to the muddy bottom, while they held their rifles and precious ammunition high over their heads. There were a few grim laughs when one of the shorter men stepped into a sinkhole and disappeared—leaving only his wide-brimmed hat floating on the muck. One of the taller men quickly reached down and pulled the sputtering soldier to shallower water, and they ploughed on.

When they reached the shallows they opened fire and charged—if wading and cursing in belt-buckle-deep water can be called a charge. After routing the Filipinos and reaching the Bagbag River, the Iowans constructed a floating bamboo footbridge. The men stripped the loads from pack animals and shouldered them, hiking unsteadily across the rickety bamboo while the animals swam across the river. After scattering the remaining defenders, they marched into San Fernando.

Emil's brigade had trekked more than 200 miles through hellish terrain and fought more than thirty engagements. The cost was high. In addition to many sick with various diseases, six officers and forty-seven enlisted men were killed, with twenty-two officers and 331 enlisted men wounded.[11]

After resting a few days, the Iowans were ordered to Cavite for garrison duty, and if it was duller work than fording swamps under fire, no one complained. Sitting around the fires as the tropical sun dropped swiftly below the tree line, they had a chance to catch up on the gossip of this hastily assembled army of farm boys, leavened by a few old Indian-fighting regulars.

The campaign had its share of characters. One of the most flamboyant was Brigadier General Joseph Wheeler. Known as "Fightin' Joe," his diminutive size (he was barely five feet, five inches and weighed in at 120 pounds) belied his history as a hell-for-leather Confederate cavalryman. It was Wheeler whose rapier-like raids had frustrated, bedeviled, and raised general hell with General William Tecumseh Sherman's Army of the West. A West Point graduate, Wheeler's allegiance was to the South, and he rose to the rank of lieutenant general in the Confederate army. At one time he commanded all cavalry forces in the Army of the Mississippi, fought 400 engagements, was wounded three times, with sixteen horses shot out from under him.

When the Civil War ended, "Fightin' Joe" was elected to the U.S. House of Representatives from Alabama. When the Spanish-American War broke out, he offered his services to the U.S. army and was commissioned a major general of volunteers. After charging up San Juan Hill with Teddy Roosevelt and fighting courageously in Cuba, he was commissioned a brigadier general and took command of a brigade in the Philippines.

The sixty-one-year-old former Confederate officer proved impetuous to the point of insubordination, always charging into enemy positions regardless of orders. When his fighting spirit was curbed by assigning him to command supply lines, he accused the U.S. general commanding his division, one Arthur MacArthur, father of young Douglas MacArthur still at West Point, of keeping him out of the battle line for fear that a "Reb" general would show him up. Since MacArthur was a hero of the Union forces during the Civil War, perhaps "Fightin' Joe" had a point. One of the great campfire stories endlessly repeated by the troops was that in the excitement of battle Wheeler would ride up to the firing line screaming "Give them Yankees hell, boys."

After eight months of hide-and-seek, punctuated by intermittent skirmishes with the *ladrones,* or thieves, the 51st Iowa was scheduled to return to the United States. They had served valiantly, but not without blemish, as one soldier reported, "Talk of the natives plundering towns; I don't think they can compare with the 51st Iowa."[12] With or without plunder, the regiment departed Manila on September 23, 1899, aboard the troopship *Senator* bound for San Francisco. From there, they returned to the small towns of Iowa where both the cornfields and their corn-fed "gals" seemed like heaven after months of homesickness and hardship in the Philippine jungles.

A few, among them Emil Holmdahl, who seemed to thrive on jungle fighting, stayed behind. Conditions in Manchu-dominated China were becoming more turbulent by the day. A wandering mercenary soldier named Edmund F. English, sporting the title of "General," drifted into Manila recruiting a foreign legion of experienced soldiers. Their mission was to aid the aging Dowager Empress of China, Tzu Hsi, in putting down a series of local rebellions that threatened her rule.[13]

The "legion" was sponsored by the Chinese Empire Reform Association, which consisted of Western-oriented Chinese both in China and in California, and a group of well-heeled American businessmen. Their motives were ostensibly lofty. They wished, they said, to bring China into the new century as a modern, respected, and independent state.

They realized that: "The Chinese, through long centuries of heredity, are an unwarlike people, the Imperial government realizes that the organization and disciplining of this army must be done by foreign officers." To find them, they looked to rugged American veterans on the West Coast and in the Philippines.[14]

The regiment was to be known as the "Royal Imperial Guard, Sinim Order of Dragoons" and would function both as shock troops and royal bodyguards. What promises of gold and glory General English made to the more naive young American soldiers were never recorded, but to the teen-aged Iowa boy, soldiering in China seemed to promise an exciting adventure. Holmdahl quickly presented himself to the general with a snappy salute.

General English promptly commissioned the youngster as an ensign, possibly because of Emil's exquisite handwriting and his flair for expressing himself. Within days the general, his newly commissioned farm boy ensign, and a motley collection of discharged soldiers, wharf-rats, and European freebooters set sail for China.

By the time their ship entered the harbor at Shanghai, the situation had undergone a drastic upheaval. Hatred for all "barbarians" (anyone who was not Chinese), was at a fever pitch in the ancient kingdom. Treaties forced upon the proud Chinese by Europeans had robbed them of sovereignty over much of their own land. Their ancient gods were humiliated by caravans of Christian missionaries who sought to turn them from their age-old religions.

Maddened by the actions of the Europeans and impoverished by natural calamities when the Yellow River flooded its banks and inundated more than 5,000 miles of fertile plain, the Chinese were on the brink of total despair. Roaming shamans, soothsayers, and chieftains of growing secret societies foretold the end of the world. The sons of Han would only be saved, they said, if the European "Big Noses," whose presence had angered the gods, were driven from Chinese soil.

The hatred of the barbarians coalesced around masters of the martial arts in Shantung province. Forming a loosely organized society named *i-ho-ch'uan*, or the "Righteous and Harmonious Fists," called Boxers by the Europeans, they went on a rampage in 1900 destroying anything European. They beheaded foreigners, burned missionary outposts and looted European trading centers. The Boxers forced the Manchu Empress to sever any ties with the "big noses with white faces." These events left General English and his doughty band of warriors without a paymaster, and the Sinim Order of Dragoons quickly disintegrated into a bunch of drifters in Shanghai. Fortunately, young Emil had the price of a return ticket to Manila.[15]

Arriving back in the Philippines in mid-1900, he found that the hard-pressed regulars were offering a $500 bonus to those signing up for more war, so Emil enlisted in the 20th United States Infantry Regiment. If going from an ensign in the Royal Imperial Guard of the Empress of China to a still underage private in the U.S. Army bothered the young soldier, he kept it to himself.

There were some compensations for joining the regulars. Paydays were guaranteed and, most important, the soldier never again had to fight with the rifle issued to volunteer regiments, the antiquated, faithful, single-shot Springfield .45-70, which kicked like a mule. Worse, its cartridge used black gunpowder which when fired sent up a pillar of smoke that gave away the shooter's position.

As regulars, the 20th was issued the modern, Danish-designed Krag-Jorgensen bolt-action rifle with a five-shot magazine. It fired the new high-velocity .30-40 caliber cartridge and, best of all, it used smokeless powder. It had a range of 2,000 yards, although it took an expert shot to hit a man at 1,000. Emil quickly qualified as one of those expert marksmen who could take out a guerrilla at that range.[16] And no doubt, he joined with his comrades in singing the little ditty that began:

> *Damn, damn, damn the Filipinos,*
> *Underneath the starry flag,*
> *Civilize 'em with a Krag.*[17]

It was to become more than a lyric—it became a reality.

The 20th U.S. Infantry was a big, tough regiment of regulars who had won a multitude of battle honors during the Civil War and had recently fought gallantly in Cuba. They more than lived up to their regimental motto, *Tant Que Je Puis* (To the limit of our ability). In November 1899, their muster rolls included forty-three officers and 1,478 enlisted men.[18]

When they first arrived in Manila, the 20th was not given a fighting assignment; its soldiers instead were detailed as military police. The army, growing distressed by a series of drunken incidents and an increasing rate of venereal disease, thanks to the multiplying number of bars and whorehouses in that city, set a curfew of 7:00 p.m. It became the regiment's unpleasant duty to sweep the streets of raucous soldiers every twilight.[19]

In 1899, before Holmdahl returned from China, the regiment finally got into the fighting south of Manila in operations along the Pasig River. They became, however, both saddened and infuriated when their much admired General Henry W. Lawton was killed in late December at the town of San Mateo, eighteen miles from Manila. Lawton, in the front lines of the fighting, was dressed in his bright white pith helmet and his iridescent yellow slicker. His six-foot, four-inch frame made a tempting target, and while he was surveying the fighting through binoculars, a rebel bullet struck him in the chest, killing him instantly.

When young Holmdahl joined the regiment, they were still maddened by Lawton's death and had a simmering hatred for the Filipinos. They were fighting in Luzon when Aguinaldo was captured in late March 1901. While much of the force of the revolution was broken by his capture, there was still heavy fighting to be done and on April 5, 1901, the regiment fought a hot skirmish near Salsona, Luzon.

In November 1901, in a command shakeup, Emil and his new comrades of the 20th were put under the command of General J. Franklin Bell, with orders to pacify any insurgents holding out in southwestern Luzon.

The guerrilla war, like that in Vietnam, had become deadly. Terrible atrocities were committed on both sides, and if young Holmdahl in later years was referred to as a callous killer, perhaps his inhibitions against shedding blood came from that youthful service in Luzon. Not untypical was an order given by Brigadier General Jacob W. "Hellroarin' Jake" Smith, who instructed a subordinate commander to take no prisoners and kill anyone capable of bearing arms. Anyone over the age of ten was old enough.[20]

Ley Fuga—meaning law of the fugitive—which authorized captors to shoot escaping prisoners was very much in vogue. There was a very bad joke about some "Tennessee Boys" who were ordered to take thirty wounded guerrillas back to an American hospital. They passed through a prosperous village, and when they finally arrived at the hospital they had "a hundred chickens and no patients."[21]

With misguided Yankee ingenuity, the troops developed an effective method of dealing with stubborn prisoners who refused to reveal military information. Quaintly named the "water cure," it was administered by stretching a prisoner on his back, prying open his mouth with a stick or bayonet, and pouring large amounts of dirty water down his throat until his stomach blew up like a balloon. When the "patient" was full, one of the "doctors" would kneel or stomp on his stomach until the water was expelled through a variety of orifices. After the "cure," most surviving prisoners were more than cooperative about answering questions.

For variety, there was also the "rope cure," consisting of wrapping rope around the prisoner's neck and body a number of times until it formed a sort of a girdle. A stick was placed between the ropes and twisted until it smothered and garroted the victim. It too was an effective cure for silence.[22]

The Filipinos, however, were not without blemish, and when their bolo-men ambushed an American, not content to merely kill him, they often dismembered the body and carved it into small pieces. As a mark of guerrilla humor, dead Americans were propped up with their throats slit and their penises stuck in their mouths. In other instances, American soldiers were found buried up to their necks in red-ant beds. The ants considered eyeballs a delicacy.

It was in this tortured environment that young Emil spent his adolescence. Under those fighting conditions, one either went insane or became very, very, tough. Emil Holmdahl became very, very, tough.

Filipinos maintained that American politicians in favor of annexation deceived the American people into believing that only a small faction of the native population supported the insurrection. As the war dragged on, however, they were fond of quoting General Arthur MacArthur:

> *When I first started in against these rebels, I believed Aguinaldo's troops represented only a fraction . . . I did not like to believe that the whole population of Luzon . . . was opposed to us . . . I have been reluctantly compelled to believe that the Filipino masses are loyal to Aguinaldo and the government which he leads.*[23]

They also quoted a dispatch from General A.R. Chaffee that, "The insurrectionary force keeping up the struggle . . . could exist and maintain itself only through the connivance and knowledge of practically all the inhabitants."[24]

In an effort to moderate what had become a very nasty war, 325-pound William Howard Taft was sent to the Philippines as high commissioner. He affectionately referred to his Philippine charges as "Our little brown brothers." The infantry out in the *barrios* responded with a ditty:

> *They say I've got brown brothers here*
> *But still I draw the line*
> *He may be a brother of Big Bill Taft*
> *But the son of a bitch ain't no brother of mine.*[25]

Officially the insurrection ended July 4, 1902, although intermittent fighting continued for another decade. When the final count was made, "benevolent assimilation" had resulted in the deaths of more than 4,000 American soldiers, almost 3,000 wounded, and large numbers enfeebled for life from tropical diseases. The insurgent army had an estimated 20,000 killed in combat. Worst of all, 200,000 Filipinos died either of famine brought on by the war or from outrages committed by soldiers on both sides.[26]

Emil Holmdahl (standing, far left) and his comrades in the 20th Infantry Regiment during the Philippine Insurrection, c. 1900. — Holmdahl Papers

⇒ 2 ⇐
Fighting The Moros

It is idle to suppose that the Moros can be subdued and made into decent citizens by throwing kisses at them.
— Dean C. Worcester, Member of
the Philippine Commission, 1900-1913.[1]

Following the official end of the insurrection, some of the nastiest fighting in the Philippines began on the southernmost islands in the Sulu Sea. In a twelve-year campaign, from 1903 to 1915, the thinly-spread U.S. Army, aided by a newly-recruited force of Filipinos trained in scout companies, attempted to pacify the factious, bandit-ridden peoples known as Moros.

Inhabiting the huge southern island of Mindanao and the smaller islands stretching to the southwest through the Sulu Sea, the Moros raised havoc with other Filipinos, raiding, looting, murdering and taking slaves. For two and a half centuries the Spanish had been unable to control them and generally, when possible, left them alone. When the Spanish abandoned the principal island of Jolo after the peace treaty was signed, the last vestige of law and order vanished. The Moro chiefs took it as a license to raise havoc throughout hundreds of islands, ranging from Mindanao to Borneo.

The Moros numbered about 275,000 in the early 1900s. Their past successes in fighting the Spaniards had led them to develop a belligerent contempt for all Christians. To add to their fanaticism, they suspected the Americans were planning to convert them by force to Christianity. In less than three years American troops fought more than one hundred engagements against them, and Corporal Holmdahl saw his share of combat.

On August 6, 1903, Major General Leonard Wood was appointed military governor of the islands dominated by the ferocious tribe. His jurisdiction covered most of Mindanao, Palawan, Bisalin, Jolo, and several hundred lesser islands of the Sulu Archipelago. Wood was both brilliant and controversial. He joined the army in 1886 as a surgeon in the medical corps. During the Spanish-American War, he commanded Theodore Roosevelt's famous Rough Rider regiment. After combat in Cuba, he was appointed military governor of that island and promoted to brigadier general.

A man of extraordinary energy and driving ambition, he immediately determined to put an end to the Sulu Sea arena of looting, piracy, slave raids, localized clan warfare, and general disregard for law and order. In one of his first reports to the U.S. War Department, he stated he would bring "order out of the chaos existing among these savage peoples."[2] With this he lit the fuse for a bloody series of fights between the Americans and the Moro clans.

At first General Wood thought the pacification would be easy and wrote to his friend, Theodore Roosevelt, now president of the United States, "One clean-cut lesson will be quite sufficient for them." They are, he wrote, "religious and moral degenerates."[3] To beef up his forces in the Sulu area, the 20th Infantry was moved from Fort McKinley on Luzon to a post at Zamboanga on the southwest coast of Mindanao. As lawlessness increased on some of the outer islands, Holmdahl's I Company was ordered to Jolo, a small island in the Sulu Archipelago, which was a hotbed of muslim fanaticism.

As the troopship carrying I Company sailed through the emerald-green seas off the Sulu Archipelago, from its deck Holmdahl could see neighboring islands whose mountains were covered with lush green tropical vines. It must have seemed to a young and romantic soldier like Holmdahl that they were journeying into a peaceful tropical paradise.

Unknown to most of the troops, however, was that the sleepy lagoons of these colorful islands harbored the swift craft of fierce pirates. The small villages built out on pilings were the haunts of gun-runners, who supplied the wants of Moro bandits. Emil and his comrades were soon to sample some of the merchandise.

Jolo was inhabited by fanatical muslims, many of whom took solemn vows to die taking the blood of Christians. With the blood of a Christian still warm on their hands, a dead Moro warrior would immediately fly to the muslim heaven "on a white horse with a green mane. He will there be washed, fed and waited upon by fifteen or twenty women forever."[4]

Moros were of medium height, physically robust, and totally without fear. They had a passion for the gaudy; the men wore tight-fitting pantaloons, colorful shirts and jackets, topped off with a bright turban wound several times around the head. From their ears hung tinkling earrings of metal and seashells.

Their traditional weapons were a razor-sharp, two-handed sword known as a *barong* that could cleave a head in a stroke. For thrusting, they used the deadly *kris*, a straight stiletto-type knife notable for its wavy steel blade. Sometimes they wore coats of mail crafted from metal wire and buffalo horn.

According to Major-General Hugh Scott,

> *In 1903 the Moros were well armed with Remington rifles, Sniders, muzzle-loaders, spears, barongs, and lantakas or bronze swivel-guns of two to three-inch caliber of native manufacture. The Remington rifles were obtained in part from the Spaniards . . .*

The Snider rifles were smuggled in from North Borneo. Worse, when Spanish soldiers dumped cases of Remington ammunition into the Sulu Sea to keep them from being captured by the Americans, Moro pearl divers recovered every box.[5]

Christians, particularly Americans, learned to be wary anywhere the clan known as *Juramentados* lurked, for at any time they were apt to run amok, charge, and decapitate a hapless soldier or Filipino civilian. *Juramentado* was a corruption of a Spanish verb meaning to swear an oath, and clan members swore to die while taking the life of a Christian. When preparing for killing, they shaved their eyebrows and bound their limbs with strong vines so that they would not easily bleed to death if wounded. They dressed themselves in a garment of pure white, received a blessing from a muslim holy man, and went out searching for a victim.⁶

After I Company arrived at the island, they were stationed in the main city also called Jolo. This was a former Spanish provincial center where nearby pearl fisheries and sugar plantations were the basis of the economy. Jolo was a town of lovely broad avenues, lined with tall shade trees which blocked the blazing tropical daylight. Toward evening, little shops selling shark fins, pearls, trinkets, love potions, and wood carvings were lighted by small swaying lamps. Jolo seemed peaceful and friendly, but could suddenly turn murderous if a Moro ran amok.

On one occasion a *Juramentado* slipped through a sewer into downtown Jolo and sat down to wait for a Christian to come his way. When two American soldiers sauntered by, he let out a yell, and foaming at the mouth and waving his bolo, charged them. The two soldiers ran into a nearby billiard saloon and in a tableau that would have been more Mack Sennett than tragic, if lives had not been stake, the *Juramentado* chased the terrified soldiers around a large billiard table. Finally the two ran out into the street, pulled their revolvers, and pumped a dozen slugs into the fanatic before he fell dead.

For all his youth, Holmdahl quickly developed the survival habits of the veteran soldier. He walked with his eyes sweeping

Sgt. Holmdahl poses in his full dress uniform while stationed in the Philippines.
— Holmdahl Papers

both sides of the street and craned his neck to check for anyone coming from behind. It was a habit he retained all his life; indeed, it probably accounted for his reaching old age.

General Scott later wrote,

> *The Moro appears to have a nervous system differing from that of a white man, for he carries lead like a grizzly bear and keeps coming on after being shot again and again. The only weapon that seems adequate to melt him immediately in his tracks is a 12-gauge pump-gun loaded with buckshot. One Moro of Jolo was shot through the body by seven army revolver bullets, yet kept coming on with enough vitality and force to shear off the leg of an engineer soldier, more smoothly than it could have been taken off by a surgeon."*[7]

This was the foe that young Holmdahl and his comrades were to face for the next three years.

In a series of short, brutal battles, Holmdahl fought alongside junior officers who were to make a mark that led them to higher commands. Among them were Captain John J. Pershing and Major Hugh Scott, both to play prominent roles a decade later in the revolution that consumed Mexico. If these two eminent soldiers did not know the newly-promoted Corporal Holmdahl during those campaigns, their common experiences may later have proved a valuable bond that helped keep the young rover out of a federal prison.[8] It was Major Scott who in September 1905, signed at Jolo a certification declaring Corporal Holmdahl as proficient in the "Infantry Drill Regulations of 1904."[9]

In the southern islands, *Juramentado* fanatics were attacking American soldiers everywhere. From the 36,000 square miles of mountains and jungles of Mindanao and the volcanic wastes of Jolo, they would strike and then fade away into the wild country. Then they would regroup and prepare to attack again. The Moro fanaticism struck close to home when four of Holmdahl's buddies in the 20th were ambushed at the small village of Talai. Sergeant John McDermott was killed, his head almost severed from his body. Private Peter Vasey was severely wounded with a bolo slash across his back. The fact that the other two soldiers killed the attacker was small consolation.[10]

When American troops retaliated against Moro attacks by sending out search-and-destroy missions, the tribesmen fled their villages and barricaded themselves in stone-and-earth forts called *cottas*. As American troops approached, they rushed to and fro across their battlements, waving flags and beating gongs until the Americans came into range. Then they blasted away with their ancient muzzle-loading *Lantacas* cannon. Screaming curses and waving *bolos*, they climbed over their battlements and charged, hoping to get close enough to lop off a few infidel heads.

After a few such encounters, American commanders used more methodical, if not quite so dramatic, tactics. Staying out of range of the Moros' obsolete cannon, they deployed modern field artillery batteries to blast the *cottas*. Then they followed up by raking the smashed forts with Maxim machine-gun fire before sending in the infantry to mop up any surviving tribesmen. It was during these campaigns that Holmdahl gained his expertise with machine guns.

Treachery was not an unknown custom among the Moros. Following a battle in Jolo, a chief named Panglima Hassan was captured. As he was escorted to jail by troops under the command of Major Hugh Scott, the chief turned to Scott and begged to be allowed to see his family, hiding in a nearby house. Scott, moved by compassion, agreed and led the group to the dwelling.

Suddenly the door flew open, and a "family" of a dozen armed Moros rushed out shooting. Scott was shot in both hands, losing one finger on his left hand and two on his right. During the melee, Hassan and his men escaped by running into the brush. Major Scott spent four months in an army hospital in Manila pondering his folly.[11]

In April 1905, Datto Pala, the chief of Sulu Island, invaded Jolo with a large force, and General Wood, with the 20th Infantry and other troops, marched out to meet him. Eight miles from the village of Mayhbun, near a small lake, the Moros dug rifle pits, set up trip wires, and prancing and yelling along their embankments, defiantly invited the Americans to attack. They got their wish.

After three days of shelling the *cottas* and machine-gunning their trenches, General Wood gave the order to advance. With a shout, Corporal Holmdahl and his comrades of the 20th went in with bayonets fixed. After vicious hand-to-hand fighting against *bolos* and *krises*, they overran the Moro entrenchments.

During the fighting, seven Americans were killed and twenty wounded; Datto Pala and 250 of his men were transported into muslim paradise by Holmdahl and his comrades.[12] Moro casualties were always frightfully high in these engagements. Anti-colonial forces in the United States were quick to accuse American soldiers of practicing genocide, since women and children were often among the native casualties.

Typical of the criticism was a speech delivered by George Frisbie Hoar, Republican senator from Massachusetts, who on the Senate floor fulminated, "You (the American army) make the American flag in the eyes of a numerous people the emblem of sacrilege in Christian churches, and of the burning of human dwellings, and of the horror of the water torture." He further accused the army of "arson, pillage and torture."[13] It wasn't always a fair criticism.

Captain Frank Ross McCoy, General Wood's aide-de-camp, wrote after the storming of a Moro camp:

> *It was most remarkable the fierce dying of the Moros. At every cotta efforts were made to get them to surrender or to send out their women and children. But for an answer a rush of shrieking men and women would come out cutting the air with bolos and dash among our soldiers like mad dogs. We had no choice, they were wiped out.* [14]

An article in the *Army News* published for soldiers stationed at the Presidio at Monterrey, California, quoted Brigadier General Frederick D. Grant in defense of the army in the Moro campaigns:

> *The Moros are as near barbarous as any tribe in the Philippines . . . They believe death at the hands of a Christian only brings heaven, quicker and more beautiful . . . They are never conquered until dead . . . Their women fight as fiercely as the men, and they can wield a kris with quickness and strength.*

As Grant explained:

> *There are no distinguishing features in dress and a soldier fighting for his life is not apt to hold his fire to determine which are men and which are women. If he did, it would mean an army funeral.*

The *Army News* further quoted a Manila newspaper that reported that in one fight, "The Moros, both men and women, assailed the Americans with hand grenades containing nails, bullets and spear heads."[15]

The Moros had another trick which resulted in American casualties. Before a tactical retreat they covered their "dead" with white shrouds and left them in their vacated trenches. When Holmdahl and his comrades leaped into the abandoned trenches and made the mistake of turning their backs to the deceased enemy, "...the 'dead' suddenly sprang to life and plunged a *kris* into their backs."[16]

The fighting on Jolo was part of the Third Sulu Expedition which lasted through May 1905. Now twenty-one-years-old and a hardened veteran, Holmdahl fought in three other engagements at Tambang Market, Ipal, and Palas Cotta. In his service record was placed a note stating, "This soldier has military ability and zeal to fit him for a commission as an officer in a unit of United States volunteers."

On December 15, 1905, Holmdahl was appointed Sergeant of I Company, 20th Infantry Regiment. He was twenty-two-years-old, and in a regular army regiment composed of veteran soldiers, he was very young to hold that rank.[17] In March 1906, the 20th Infantry was rotated back to the United States. After more than a month on the troopship *U.S.S. Sheridan*, Holmdahl arrived in San Francisco harbor. An army band met them at the dock playing "Home Sweet Home" and "When Johnny Comes Marching Home."[18]

U.S. soldiers on guard duty amid the ruins of San Francisco, 1906. Holmdahl is standing, second from the right. — Holmdahl Papers

3
Earthquake in Old 'Frisco

*"Not in history has a modern imperial city been so
completely destroyed. San Francisco is gone!
Nothing remains of it but memories..."*
— Jack London

After eight years of soldiering in the Philippines, Sergeant Holmdahl walked down the gangplank onto the San Francisco Embarcadero on March 6, 1906. Tall, rawboned, tanned, and muscular, he bore little resemblance to the apple-cheeked farm boy who first shipped out to the Orient with dreams of adventure. The biggest change, perhaps, was about the eyes; no longer the wide-eyed innocent, his cold blue irises looked out with a perpetual squint caused by exposure to the tropical sun. When Holmdahl faced danger, those eyes contained all the warmth of a pair of blue ice cubes.

His mouth reflected a sardonic smile. For all this, he radiated a cool charm, was engaging to women, and slightly frightening to men. The noted American soldier and military historian Brigadier General S.L.A. Marshall, wrote, "He was the most handsome man I have ever seen."[1] He was a spitting image of the movie star Clint Eastwood, and if he was a man who had killed, he was not necessarily a man who enjoyed killing, as one accuser later described him. But he certainly had no compunction about taking a life if it was necessary or even convenient. Years of service under the likes of General Jake "Hell Roarin'" Smith had removed those inhibitions.

After disembarking in San Francisco, the 20th was transported to their new post at the Presidio at Monterey. With a long leave accumulated over his years of service, Holmdahl relaxed in the sybaritic pleasures of San Francisco's notorious Barbary Coast. After "painting the town red" the night before, he lay stretched out on his hotel bed the morning of April 18.[2] Perhaps the previous night he had listened to the greatest tenor of his time, Enrico Caruso, sing an impassioned Don José in <u>Carmen</u> at the Grand Opera House. More likely, he had hob-nobbed at Barbary Coast bars, frequented that evening by the young actor, John Barrymore.

But suddenly, there was a wrenching shudder and the walls of the hotel began quivering like one of the Barbary's exotic dancers. A picture flew off the wall and the ceramic pitcher slid off the dresser, crashing to the floor. The lanky Holmdahl was flung out of bed as the room turned topsy-turvy. Clutching his uniform, Holmdahl dashed from the building clad only in his shorts, as the hotel shook to pieces. His welcome back from the Orient was the famed San Francisco earthquake, the greatest disaster ever to hit the West Coast.

Staggering down the street that morning he saw flames shooting high into the air. There was a deafening roar as 28,000 buildings collapsed and burned, and hundreds of lives were lost, as more than four square miles of the city were reduced to rubble. Perhaps it was then that Holmdahl became convinced that he could survive anything.

At the plush Palace Hotel, Caruso, "waked up, feeling my bed rocking as though I am in a ship . . . I run into the street. That night I sleep on the hard ground." Never again, never, Caruso told his friends, would he ever sing in San Francisco again. And he never did.[3]

Writing to his sister, Barrymore reported he had been thrown out of bed by the tremor and had wandered "dazedly" into the street. How much was earthquake and how much was hangover is unknown; he did not elaborate. Spotting the lurching actor, an Army sergeant, impressing work details, put a shovel in Barrymore's hands and ordered him to shovel debris for twenty-four hours. When this was relayed to his uncle, the equally famous actor, John

Drew, he commented in wonderment, "It took an act of God to get him out of bed, and the United States Army to put him to work."[4] During the early hours of the disaster, the army was somewhat hard-nosed about selecting "volunteers."

The *Army News* of April 26, 1906, recounted the experiences of a young man who was strolling down an avenue seemingly unconcerned about the ruins about him. He was fashionably dressed in "a summer suit, straw hat and kid gloves." This "Adonis," the paper reported, was

> *... grabbed and ordered to help clear the bricks and other debris from the trolley car tracks. At first he hesitated, but the sharp point of a bayonet convinced him ... for the next five hours he was doing a laborer's work in spite of his handsome attire.*[5]

In downtown San Francisco, Charles F. Curry, Secretary of State for California, was grabbed by a military detail cleaning up Market Street below the site of the Palace Hotel. Suddenly he was spun around and a shovel was thrust into his well-manicured hands, although he protested he was an official on important state business. A tough-looking corporal stuck his face up against the face of the Secretary of State for California and snarled, "I don't give a damn who you are. Start heaving those bricks." For the next hour and a half, Secretary Curry shoveled bricks and other debris into carts, until a ranking army officer recognized him and relieved him from his "volunteer" work detail.[6]

Opposite the ruins of the city hall a husky sergeant had a squad of fifty citizens pitching bricks out of the middle of the street. "Ain't they doing fine?" said he with a grin:

> *I've got the chief of police from Milpitas or somewhere in there throwing brick. He told me who he was but I persuaded him. He's doing well. We'll have this street open clear to the ferry before night. See if we don't.*[7]

Picking through the debris, Holmdahl found transportation down the coast to Monterey, where he reported for duty at the Army post. All in all, it had been one hell of a leave. After reporting in, Holmdahl and most of the army troops in California were quickly transported to the stricken city where they faced scenes of incredible desolation.

In the words of writer Jack London, "San Francisco is gone. Nothing remains of it but memories . . ."[8] Hundreds died in collapsing and burning homes, and more than 225,000 people who survived the quake and the fires were homeless. The streets were piled high with the debris of proud buildings; telegraph and telephone wires were strewn about like spilled spaghetti; there was no gas or electric power. The city was in darkness except for the flames of burning buildings which created a fire storm feeding on twisted piles of wreckage. Destitute men, women, and children with ravaged faces, carrying what possessions they could save, fled from the flames.

Dedicated police, firemen, and soldiers finally brought order to the panic-stricken population. The troops were commanded by General Frederick Funston, the man who accompanied by only a few volunteers had probed deep into insurgent territory in Luzon and captured the Filipino rebel leader Emilio Aguinaldo. After water mains burst and pressure died, hard-worked fire engines became useless. Setting backfires and dynamiting buildings to build fire walls, Funston's men slowly brought the conflagration under control. As usual, in the wake of such horrors, human hyenas began prowling the streets pillaging and looting private homes of any valuables spared by the fires.

But, Mayor E.E. Schmitz was made of sterner stuff than many later mayors. He proclaimed:

> *The Federal troops, the members of the Regular Police Force, and all Special Police Officers have been authorized to KILL any and all persons found engaged in looting or in the commission of any other crime.* [9]

Some of the wharf rats and other denizens of the notorious Barbary Coast made a fatal mistake in not taking the warning seriously. As the pillagers slunk down battered streets, Holmdahl and other veteran soldiers, accustomed to hunting Filipino guerrillas in the dark, picked looters off by the dozen. While some of the fainthearted complained about the soldiers' "brutality," the looting quickly stopped.

Holmdahl later wrote, "It [San Francisco] was worse than soldiering in the Philippine Islands. I was on guard at the United States Sub Treasury Building for 125 hours with little sleep."[10] During that time someone took a photograph of Sergeant Holmdahl at his post showing smashed and burnt out buildings in the background. (see page 28) After services were restored and the city was capable of being controlled by civil authorities, the troops were withdrawn. Holmdahl's unit, exhausted by nights of standing guard and dueling with vandals, was ordered back to Monterey. Sadly, there was not much left of the city, and Jack London reported, "San Francisco, at the present time, is like the crater of a volcano, around which are camped tens of thousands of refugees."[11]

At Monterey, Holmdahl performed routine and humdrum duty, and much of his free time seemed to have been spent playing baseball. On Thanksgiving Day, the *Army News* reported, he played second base on a team of enlisted men that defeated the officers 12-2 before all adjourned to the mess hall for a dinner of roast turkey. The *Army News* of December 6, 1906, lists Holmdahl as the winning pitcher on the Company "I" baseball team which defeated rival Company "H." On December 13, that newspaper reported he pitched on a team that defeated Monterey High School.[12]

Holmdahl received his discharge on January 31, 1907. Footloose, the ex-sergeant had a number of skills greatly in demand during the first years of the twentieth century, for this was the heroic time of the European and American mercenary soldier. In those days the rugged military adventurer was romantically described in prose and poetry by newspaper correspondent Richard Harding Davis and other "yellow journalists" of the day.

During those years the natives were restless throughout the banana republics of Central America. A man who not only knew jungle fighting tactics, but also had the technical skills to fire and, most important, maintain the new-fangled machine guns could command a fine price with the sprouting revolutionaries of the day. To the young ex-sergeant, the lands of the bananas seemed to offer new opportunities for adventure and riches.

Map 2: *Central America c. 1900.* — Provided courtesty of The General Libraries, The University of Texas at Austin

4
The Banana Men

"But there is in our fellow citizens an innate martial spirit, a yearning for adventure, which finds vent whenever any expedition, with the prospect of a fight with any one at all, is undertaken . . . the United States is filled with soldiers of fortune."
— Arthur H. Dutton

After receiving his discharge, Holmdahl worked a few weeks in Oakland as a steam fitter, then a cryptic note in his handwritten diary states, "In the vaudeville business, Imitation Animal Act greatest hit on the Orpheum Circuit."[1] Perhaps a phony animal act was the genesis of Holmdahl's later talents as a confidence man. Then, there is a curious gap of two years in his diary entries. Consequently, it is not known how or when Holmdahl got involved in the so-called "Banana Wars," waged intermittently in Central America from 1880 until 1930.

Holmdahl later told friends of his fighting experiences under General Lee Christmas, perhaps the greatest mercenary of them all; but no written records exist. Undoubtedly, he was there because many of the "banana men" who fought in those jungles also played a prominent part in the Mexican Revolution and knew Holmdahl, if not in person, at least by reputation.

The opposing sides in that struggle often resembled participants in a game of musical chairs (although losers were not only out of the game; they were probably shot). An old Central American comrade of today might turn up on the other side tomorrow. Since no one ever implied Holmdahl was a fraud, lack of records notwithstanding, one may assume he took part in at least some of the many revolts and wars in Honduras and Nicaragua from 1907 until the summer of 1909. While it is unknown exactly what he did, some probable assumptions can be made by following other mercenaries in the jungle wars of Central America.

San Francisco and New Orleans were the prime recruiting areas for the hard-faced men who fought for any side that paid in cash. There was little, if any, ideological content in those revolutions, and although one side might call themselves "liberal" and the other side "conservative," the labels meant little. It was always a battle between a family dynasty of *ricos* who were "in" and another dynasty who were "out." Often an American banana company, wanting a land concession or lower taxes or to wipe out a competitor, would finance a revolt and recruit and pay the hired guns.

The revolutions meant little to the impoverished peasants who labored for pittances on the banana and coffee plantations. As Richard Harding Davis wrote, "Half of the people in the country will not know of it (the revolution) until it has been put down or succeeded."[2] He might have added that they couldn't have cared less. After all, it amounted only to a change of masters, each totally uncaring about their welfare. In recruiting locals to fight, the standard joke was about the Honduran agent who dispatched his "volunteers" to a rebel general with the note, "I send you 40 volunteers. Please return the ropes."[3]

As the Philippine insurrection began to falter, there were scores of ex-soldiers, looking for work, who hung out in the Vieux Carré in New Orleans, usually at the Hotel Monteleone. In San Francisco you could usually find them in the rebuilt Tenderloin District, drinking and eyeballing the girls. At those haunts they would be contacted by entrepreneurs like Samuel "Sam the Banana Man" Zemurray, the brains of the expanding American-owned United Fruit Company.[4]

After agreeing on wages, the Americans usually boarded a banana boat out of New Orleans bound for the Caribbean port of Bluefields, Nicaragua. From there, they gathered at the Hotel

Tropical bar until they moved out through tick-infested mountains and malaria-ridden jungles to their new "army" in El Salvador, Honduras, Guatemala, or Nicaragua.

While some of the men who fought in those wars were professional soldiers, others were cutthroats, thieves, or ne'er-do-wells on the run from police or from intolerable personal problems. The real professionals, the Sam Drebens, the Tracy Richardsons, Guy Maloneys, and Emil Holmdahls, while not adverse to acquiring a large stake for their services, were often driven by other devils.

Maloney once said, in the long run, he would have earned more money digging ditches, "but there was also the lure of a romantic cause, for adventure, for fame, or for the hope of future material rewards."[5] Victor Gordon, who led mercenaries in a dozen Latin American revolutions, described his recruits:

> *Some were cutthroats who fled the United States and were going under assumed names. Others were youngsters who had come down ... for excitement. A few were experienced soldiers.*[6] *Central Americans of all persuasions usually referred to them as "scum of all nations."*[7]

Often, however, the mercenaries were cheated out of their wages. A Captain Linderfeldt who led a unit of Americans in the battle for Juárez, Mexico, in 1911, complained he was promised $10 a day in gold, but was finally paid off with only $2 a day in pesos.[8] And if one picked the losing side, he was lucky to escape with his life.

Mostly, the Americans were adventurous types who had the restlessness of the now-vanished frontier still running in their veins. When asked why he joined a band of Nicaraguan revolutionaries, Tracy Richardson replied, "For money and the hell of it."[9]

These men were in contrast to foreign volunteers fighting for the Boers during their war with the British. War correspondent Richard Harding Davis said of them, "They were not soldiers of fortune, for the soldier of fortune fights for gain. These men receive no pay, no emolument, no reward." For them, Davis said, "There were no bugles ... Their conscience was their bugle call."

They fought, "to try and save the independence of a free people."[10] It was a sense of morality totally lacking in the Banana Wars.

After picking the winning side in one Nicaraguan revolution, Richardson and Sam Dreben each received $5,000 and a parcel of good banana-growing land.[11] Not being of banana-growing temperament they pocketed the cash and sailed for New Orleans. This was good pay for 1909. But when one considers that they sweated in fever-ridden jungles for months, ate rotten food, drank bad water, were constantly chewed on by a large variety of airborne, disease-bearing insects, risked being crippled or killed in the fighting and faced a lonely death in a savage land—it is doubtful the banana men fought for money alone.

Not infrequently the Central American *condottiere* would find themselves in an army commanded by the legendary General Lee Christmas. Christmas, a Louisiana railroad man, arrived in Honduras in 1894 as an engine driver for the local railroad. He soon became embroiled in a local revolution. By sheer guts bordering on foolhardiness, a native ability to lead tough men, and a determination and focus rare in Central America, he rapidly rose to prominence as a commander who could take a town and rout an enemy army.

At about the time that Holmdahl became involved, Honduras was undergoing one of its perennial revolts led by Christmas against the current tyrant. According to a newspaper interview given in 1913, Holmdahl became a field officer in that army. In that and subsequent campaigns, for almost two years he underwent a "procession of marches through sweltering jungles, descents on startled adobe towns, confused fights between rival groups of barefooted soldiers and ambuscades in lonely valleys with outlandish names."[12]

Since Honduras had no extradition treaties, it was a mecca for cutthroats from both the United States and Europe. It became a fertile recruiting ground for soldiers of fortune, con men down on their luck, and the kind of human flotsam who would join any enterprise if the price was right. Christmas's toughness was an ongoing legend among the hard cases in his army. The story was told that once when captured by the "other" side, he was brought before the enemy general who announced:

"Christmas, you are a no-good son-of-a-bitch and I'm going to have you shot."

Christmas reportedly replied, "Okay, but I have just one request before you shoot me."

"What is it?" the general growled.

"Don't bury me," Christmas said.

A little taken back, the general asked, "Why in the name of God, don't you wish to be buried?"

Christmas drew himself up to his full six-foot plus height and pointed to a nearby clump of trees, "See those filthy vultures sitting in those trees? I want them to eat my body so they will fly over your camp and dump shit on your heads."

The general looked startled and for a moment there was a deathly silence until suddenly he let out a tremendous roar of laughter. Tears running down his cheeks he chortled, "I can't shoot anyone with a sense of humor like that. Let the bastard go free."[13] Christmas was released and his legend grew.

But for all his toughness, the railroader-turned-general also needed skilled men. The age of revolution in Central America coincided with mechanical improvements of the rapid firing weapons recently labeled the "machine gun." The old multi-barreled Gatling gun required a strong man who could crank the heavy revolving barrels of the gun, regulating the rate of fire. It was extremely heavy, difficult to transport, and often jammed. The new machine guns such as those designed by Sir Hiram Maxim, and similar weapons being produced by British, German, and French manufacturers were lighter and had a much more rapid rate of fire. Strong arms were no longer needed because to get the maximum rate of fire from its one barrel, one had only to pull the trigger.

Early machine guns, however, were temperamental and tended to jam. To keep them operating in a jungle environment was beyond the capacity of the average peasant soldier. The peasants were brave enough, but it was said they had the ability to break an anvil and could repair nothing. It was a golden opportunity for all the footloose veterans of the Spanish-American War and the Philippine Insurrection, as well as a number of strays from the British armies and their scores of colonial battles.

Soon they came. Sam "The Fighting Jew" Dreben, who fought in half a dozen wars and in middle age became one of the most decorated war heroes of the American Expeditionary Force in France in World War I.[14] Tracy Richardson, "The World's Greatest Machine Gunner," who single-handedly captured the city of Managua and later became an officer in the Canadian army, the British navy and the U.S. Army in World War I. He survived to serve as a Lieutenant Colonel in the U.S. Army Air Corps during the second World War.[15]

There was Guy "Machine Gun" Maloney who became a colonel and commanded batteries of artillery in the U.S. Army in France and later was named chief of police in New Orleans, Louisiana; Edward "Tex" O'Reilly who fought in Asia, Central and South America and later became a famous editor and war correspondent; and one Emil Holmdahl, who was perhaps too much the professional soldier to rate a nickname.[16]

Those men acquired local, and sometimes world-wide, notoriety as fighting men in the banana wars, thanks to the sensational "yellow journalism" of the times. Other mercenaries would find only lonely unmarked graves in a tropical jungle. If their glory was tarnished, few of these soldiers of fortune would acquire wealth from their sometimes heroic efforts with foreign armies. These were the "Banana Men," whose fighting prowess changed the course of Latin-American history.

"Tex" O'Reilly fought under eight flags and then became a famous newspaper editor in San Antonio.
— Courtesy San Antonio Light

Holmdahl likely fought alongside Christmas in a 1907 battle between an invading Nicaraguan army on one side and 2,000 Honduran troops led by the former engine driver. During that war, Tracy Richardson, fighting beside Holmdahl, embellished the famous "buzzard shit" story. In his memoirs, he states that when

Christmas was captured, he was staked to the ground, his shoes removed, and his feet burned with red-hot machetes before he was led out to be shot.[17]

Following a series of inconclusive battles, the American government called the warring parties together and established a cease-fire and shaky peace. In late 1907 and early 1908, Christmas led another revolution to install his friend Manuel Bonilla in the president's chair in Tegucigalpa, Honduras. It is likely that Holmdahl took part in that campaign as well.

After that revolt succeeded in the fall of 1909, many of the mercenaries moved to Nicaragua to join an army attempting to unseat the tyrant José Santos Zelaya. They may have had tacit approval from the United States government. Zelaya, with European and Japanese backing, was making noises about digging a canal across the Nicaraguan isthmus, which would compete with American plans for the Panama Canal.

In fighting along the San Juan River near Lake Nicaragua, Zelaya's troops captured two American mercenaries, Lee Roy Cannon and Leonard Groce. They were caught in the act of laying mines designed to blow up steamer traffic on the waterway. The Americans were tortured until they signed confessions admitting all sorts of evil deeds, and after a farce of a court martial, the hapless pair were condemned to die before a firing squad.

In spite of protests by the United States consul in Managua, the two were led from their filthy cell during a rainstorm and made to sit on a bench in front of recently dug graves. Their feet and hands were tied and ragged bandanas were placed over their eyes. Four riflemen, two to each of the condemned, stood six feet away, and at the command "*Fuego*," they shot the two and unceremoniously shoved them into the graves.

Groce left a Nicaraguan wife and four children. Cannon wrote an agonizing letter to his mother saying, "Now, mother dear, bear up. This is my fate; the results of war and disobedience to a loving mother."[18] Richard Harding Davis and other glamorizers of the mercenaries notwithstanding, the price of glory for soldiers of fortune often came high.

The deaths of the two Americans, however, probably gave William Howard Taft, now president of the United States, all the excuse he needed. He dispatched a fleet and landed U.S. marines in Nicaragua, and the two dead mercenaries suddenly became American heroes engaged in a battle for freedom and liberty.

Overwhelmed by new-found patriotism and banana company gold, a flock of new American volunteers began to swell the army of Nicaraguan revolutionists under General Luis Mena. Marching through swamps and bogs, hacking their way through interlaced jungle vines, fighting malaria and exhaustion, Mena's army with his 400 American machine gunners fought their way toward Managua. As further intervention by the American government seemed imminent, Zelaya gave up and accepted a courtesy ride on a Mexican gunboat which took him into exile. His successor found favor with the Americans, and, with their support for the rebels withdrawn, the revolutionary forces suffered several stiff defeats.

At this point many of the American mercenaries, probably including Holmdahl, considered discretion the better part of valor and boarded a banana boat bound for New Orleans. The pickings, they thought, might be better in the revolution about to break out in Mexico against the dictatorship of President Porfirio Díaz. It would, indeed, become lucrative for some; for others it would become another graveyard.

A story in the *El Paso Times* later reported that Holmdahl made his way to New Orleans and then . . .

> . . . *joined a filibuster expedition . . . for South America with a shipload of ammunition. The ship circled Cape Horn and landed its cargo at Mazanillo, in the Mexican state of Colima. The soldier of fortune then went to Los Angeles where he joined a junta planning a revolt in Mexico.*[19]

For Holmdahl, it was the first in a lifetime of hair-raising adventures in Mexico.

Map 3: *Northwestern Mexico, c. 1911* — Halcyon Press

≫ 5 ≪
Revolution in Mexico

"Compañeros del arado
Y de toda herramienta
Nomás nos queda un camino:
Agarrar un trienta-trienta!"

Comrades of the plough
And all workman's tools,
There's only one road now:
To seize a .30-.30.!
— *Corrido* of the .30-.30 carbine

In 1909 Mexico was a volcano about to erupt. For more than thirty years the people of that country had suffered under the iron fist of dictator Porfirio Díaz. Since Díaz seized power in 1876, he had done much to modernize the archaic economy of the country by encouraging foreign capital to develop its railroads, its petroleum and mining industries, and its primitive agriculture. But only a few political allies had prospered, and, as foreign interests controlled much of the economy, many Mexicans felt they were strangers in their own land.

Among the intellectual and educated classes this foreign domination set ablaze the embers of a dormant nationalism, and soon the cry of "Mexico for the Mexicans" began to be heard from provincial marketplaces to salons in Mexico City. Throughout the years of the "Porfiriato," the mass of the peasantry had lost much of their communal land to the rich *hacendados*, either from legal chicanery or from the weapons of hired *pistoleros*.

Small merchants and an embryonic middle-class smoldered with resentment, as opportunities for growth in a burgeoning economy were thwarted by the political and economic collusion of Díaz adherents which stifled their ambitions. Even among the well-to-do, some of the concepts of Jeffersonian democracy had filtered down from the northern border to offer the hope of political freedom and an escape from a dictatorship of brutal force.

There was a whiff of revolution in the air and soon it would be

> *A widespread revolution that would lead to hundreds of thousands of deaths, a revolution led by new caudillos, some modern and some archaic, divided . . . between the ideals of the future and the roots of the past.*

There would be factions of "nationalists, democrats, anarchists, socialists, Jacobins, devotees of the Virgin of Guadalupe," combining both with and against each other.[1] Into the eye of this growing political hurricane stepped a tough twenty-six-year-old soldier looking for trouble. It was not difficult to find.

In the Holmdahl Papers in the Bancroft Library there is a typed, unedited, eleven-page autobiographical manuscript that has never been published, entitled "As a Soldier of Fortune and Filibuster in Mexico." In it, Holmdahl describes his entry into the Mexican revolution, "In the year 1909, I answered an advertisement in a Los Angeles newspaper which read, 'Wanted: A man with military experience, who has nerve and is single'."[2]

Holmdahl answered the ad, listing his military experience, and was soon invited to a midnight rendezvous in a sleazy section of town. After several meetings and exhaustive interrogations, his mysterious questioners gave him an envelope containing a $100 bill and brought him before members of a revolutionary junta. They asked him if he, posing as a mining expert, "would purchase arms and ammunition and smuggle it into Mexico."[3]

If this is an improbable story, it is less improbable than many of the documented events in Holmdahl's hegiras and hair-breadth escapes in Mexico during the next twenty years. And, as will be seen, Holmdahl had reasons to conceal the origins of some of his early adventures and associations.

While Holmdahl later identifies the revolutionary junta members as part of the movement led by Francisco Madero, it is highly likely that he fell in with members of the radical Flores Magón faction. This group of plotters, led by Ricardo Flores Magón, were hard-core anarchists. They were supported in the United States by the equally radical International Workers of the World, known to Western lawmen as "Wobblies." They were the instigators of the bloody strike at an American-owned copper mine at Cananea, Sonora, in 1906, which many called the "Lexington and Concord" of Mexico.[4] In 1908 they launched an abortive attempt to topple Díaz, which was put down with much bloodshed. Operating out of California, they were busy subverting the Díaz regime a decade before the liberal revolution led by Madero. As the most radical revolutionaries, they hated liberals and later went to war against the Madero government.

During 1909, their newspaper *Regeneración* fanned the flames of hatred against the Díaz regime exhorting

> *Throw down the plough. Slaves, take the Winchester in hand . . . Work the land, but only after you have taken it into your own possession . . .*[5] *Forward comrades. Soon you will hear the first shots; soon the shout of rebellion will thunder from the throats of the oppressed . . . Land and Liberty.*[6]

If Holmdahl joined this group, he undoubtedly wished to conceal it, particularly since he later sought a commission in the United States Army. The Magonistas were constantly harassed by American officials and Magón ultimately died in an American prison.

Holmdahl recounted that he was given "plenty of money" by the group. He traveled to Nogales, Arizona, crossed the border, and boarded a Mexican train for a 500-mile southern journey to the end of the line at Culiacán, the capitol of Sinaloa. From there, he wrote,

"I purchased a horse and saddle and lit out further south . . . over the tortilla trail . . . [it got its name] as there was nothing to eat on the way except the tortillas you brought with you."[7]

After a 100-mile ride across the then sparsely populated coastal plain, he arrived at the west coast port city of Mazatlán. He spent several months there making contacts and improving the rudimentary Spanish he had picked up in the Philippines and during the Banana Wars. From Mazatlán he traveled to his revolutionary objective—Tepic, the capital of the small coastal state of Nayarit.

Founded in 1530, Tepic was a sleepy colonial village hemmed in by mountains to the east and the Pacific Ocean to the west. It was noted only for its swaying palm trees, the Church of Santa Cruz, and near the beaches some of the most voracious stinging flies and mosquitoes in the Western Hemisphere. It is in Tepic that Holmdahl first demonstrated his full blown talents as a con artist/spy and all-around dissembler.

Holmdahl in the regalia of an insurgent fighter, c. 1911 — Holmdahl Papers

Posing as a wealthy representative of a New York mining company, Holmdahl gained entree to the Governor's Palace by dangling the usual lure of the Díaz era. He told the governor he was interested in purchasing "good property" and promised "good money" would be available in exchange for official help. The governor, charmed by the suave manners and presumed wealth of the young *gringo*, invited him home for dinner. Soon, on warm evenings, they were seen strolling arm-in-arm around the town plaza as flirtatious senoritas batted their eyelashes above their fans as they passed the presentable young American. Tepic was a young man's romantic dream.

But as the revolutionary winds swept down from the mountains to Tepic, this idyll was to turn nightmarish. As Holmdahl wrote: "But now— their hearts are chilled with fear, their souls are shrunken with their pain; for death is ever stalking near."[8] It was soon to come. But Holmdahl became less sentimental for there was a job to be done.

Winning the governor's confidence, Holmdahl soon learned the number of men and the amount of arms and ammunition at the disposal of the Díaz forces in the area. The governor, according to Holmdahl, actually showed him a hiding place where he had secreted rifles for use in an emergency. The young "entrepreneur" forwarded this information to the Los Angeles junta. Unsettling at this time, however, was the arrest in Tepic of several revolutionaries, who were speedily shot in a public execution in order, as the governor said, "to put fear in the people."

Using the pretext of surveying timberland on the coast, Holmdahl spent nine days searching for potential landing places where he could bring in smuggled arms. Returning to the capital, a man approached him and whispered that the governor had been informed he was a spy, his arrest was imminent, and he should run for his life. Apprehensive, Holmdahl went to his hotel room to gather his belongings, only to see from his window a number of *rurales*, the dreaded rural police, surrounding the building.

The former young-man-about-town slipped into the hotel patio, used his leather lariat to lasso an overhanging water spout and pulled himself up to the roof. Leaping onto the roofs of nearby buildings, he reached the end of the block where lowering himself, he bumped into a dozing policeman. The policeman awoke, began shouting, and pulled his gun from its holster. He got off one wild shot before Holmdahl, firing over his shoulder, made the policeman "eat dirt."

"I knew I was in for it," Holmdahl wrote, "as I had killed this monkey." Running to the nearby stable where he kept his pride and joy, a blooded sorrel stallion which he believed could outrun any horse in the area, he swung into the saddle and rode out into the street. The *rurales*, now mounted, galloped in a frenzied pursuit as a crowd, attracted by the gunfire, filled the street leading into the plaza. Holmdahl rode into the crowd, figuring the *rurales* would not shoot into a mass of people. This was a mistake as *rurales* never gave a damn whom they shot, and immediately opened fire. "Bullets" Holmdahl wrote, "came uncomfortably close."

With his characteristic ironic sense of humor, he recounted,

> *A fat priest came out of the church and waved his hands at me. I fired, not at him, but at a large stained glass window just above his head, and shattered the glass. If you ever saw a scared fat priest make a quick retreat that 'toad' made grand time. I bet he called me a few things not in the Good Book.*

Laughing, Holmdahl doffed his sombrero as he galloped down the cobbled street. Soon, he wrote, his sorrel outdistanced his pursuers; they, however, fired a final volley, and he felt his horse stagger. The sorrel, running full out, began to weaken after several miles, and Holmdahl pulled up to a corral owned by a local rancher.

Dismounting and tugging off the saddle, he spotted a bullet hole in the horse's flank, "a few inches above the tail." Saddling one horse and leading another, he opened the corral gate and stampeded all the other horses, as he galloped down the roadway. Behind him an infuriated rancher screamed curses at the *gringo ladrone*. "It was," Holmdahl admitted, "the first horse I ever stole." After a long ride he lost his pursuers in a heavily wooded area.

After four days of hard riding, stealing horses, eating on the run, and sleeping in the saddle, he reached a hot spring near the village of Tuxpán, and, feeling drowsy, decided it was safe to take a nap. Dismounting, he hobbled his horse and fell into a deep sleep. He was torn from uneasy dreams by a sudden intense pain in his foot.

Jerking awake he saw the butt of a Mauser carbine coming down on his other foot with great force. As he looked up he was confronted by the barrels of more than a dozen rifles pointed at various parts of his body. Managing a wan smile, Holmdahl said, "Some race, huh?" The *rurale* commander responded with a hard jab to his ribs with a carbine barrel, hissing, "Your gringo sense of humor is misplaced."

Pulled to his feet, his hands were tied behind him so tightly that he complained. Annoyed, his guard stabbed him in the leg with a knife—a most effective way of quieting complaining prisoners. His captors put a rope around his neck, and for a terrified moment Holmdahl believed his career as a spy was to end with him dancing on air, suspended from the limb of a nearby tree. It was with a small sense of relief that he saw that the rope was tossed to a sergeant, who hitched it to his saddle-horn and then spurred his mount.

Half-dragged along ten miles of hot, dusty, rough road, Holmdahl was taken to the small town of Rosamorada. As he was pulled through the town, its poor inhabitants crowding the street were lectured by the *rurale* leader about the consequences of plotting against Porfirio Díaz. Holmdahl noted, however, that while the men stared blank-faced at his dust-covered, bloody body, many of the women had tears in their eyes. Exhausted, beaten, and half-strangled, he was almost glad when he was thrown into the town's foul-smelling jail.

That night a Díaz agent, "a tall, lean, mean-looking man," accompanied by a priest, entered Holmdahl's cell. A written confession was shoved in his face, and he was ordered to sign it and give them the names of his fellow traitors or he would be immediately shot. A young jailer, with a face badly scarred from smallpox, standing behind the two, caught Holmdahl's eye; he was shaking his head from side to side.

When Holmdahl refused to sign, he received stinging slaps in the face until the Díaz agent nodded to the priest to give this stubborn *gringo* the last rites. The priest asked, "What is your religion?" Holmdahl replied, "I have none and you can go to hell." The agent leaned down and slugged the prostrate man, screaming, "Respect the church." The priest, however, made the mistake of not retreating out of range, for while Holmdahl's hands were tied, his feet were free and he lashed out, kicking the clergyman in the belly. For that rather foolish act of bravado, Holmdahl got the hard steel of a carbine butt slammed against his head.

Shaken awake hours later, his head feeling like a smashed melon, Holmdahl found the guard laying a plate of frijoles and a tamale in front of him. Hands still tied, he gulped the food "hog style." Whispering, the guard said, "If you had signed they would have shot you. Now, sleep, you will need the rest."

In the early hours of the morning the guard came again. He untied Holmdahl's hands, held his finger to his lips, and motioned for the prisoner to remove his boots. Tip-toeing out of the cell, down the jail corridor they deftly stepped over a sleeping guard. They climbed the stairs and headed for the door, but as Holmdahl eased around another sleeping guard, the man stirred, opened his eyes, and reached out grabbing his ankle. His reward for his alertness was the hard heel of Holmdahl's boot slammed into his face. The prisoner and his friendly guard ran out the door into the night where two tethered horses awaited them. They leaped into the saddles and rode off into the darkness. It was to be the first but not the last time Holmdahl escaped from jail and dodged a firing squad.

After riding all night, at dawn they were camping in the mountains beside the Acaponeta River, when Holmdahl spotted a gang of laborers repairing a bridge. From the American foreman he borrowed a gun, some flour and sugar, and headed back to the security of the mountains. From several conflicting stories, as can best be determined, Holmdahl fled to the American border. He reentered Mexico in the summer of 1909.[9]

In an interview with southwestern historian Bill McGaw, in October 1962, Holmdahl said he went to work on a railroad laying track near Mazatlán, Sonora, on Mexico's west coast.[10] In a letter to the Adjutant General of the United States Army written on December 24, 1913, Holmdahl stated, "Entered Mexico on Southwest coast nine months before Madero outbreak, as spy in employ of revolutionary Junta."[11]

During that time, somehow, he obtained a commission as a captain in the Sonoran Rural Police, the dreaded *rurales*, commanded by Colonel Emiliano Kosterlitzky.[12] Frightened peasants called the colonel "The Iron Fist of Dictator Porfirio Díaz." Kosterlitzky's main assignment was to search out revolutionaries—and kill them.

All three accounts may not be in conflict, because during 1909, uprisings led by Ricardo and Enrique Flores Magón erupted in June and the disgruntled peasantry in the northern states of Chihuahua, Coahuila, and Sonora were spreading rumors of revolt. Unrest spread by the Flores Magóns was to burst into flames of revolt the following year. As one of Díaz's most trusted officers, Kosterlitzky acted as the eyes and ears of the dictator, sending reports directly to the presidential palace in Mexico City. He had developed a network of informants in every bar, bordello, and village in northwestern Mexico. These informants funneled to his headquarters any conversations, movements, or rumors that could be construed as hostile to Díaz.

Considering that Holmdahl arrived in Mexico during this critical and suspicious time, it would be surprising if he was not approached by one of Kosterlitzky's agents. With a key job on a strategic railroad, the mercenary probably became one of Kosterlitzky's spies called *zorros* or foxes, by a fearful people. Playing a double game with Kosterlitzky must have been a nerve-wracking experience for the colonel was one of the most feared men in Mexico and a sinister legend along the U.S.-Mexican border.

Born in Moscow, the son of a Cossack calvary officer, Kosterlitzky became a naval officer candidate as a teenager. Bored with navy discipline, he jumped ship in Venezuela and fought his way north to Mexico, where he joined the Mexican army and fought Apaches, Yaquis, and Mayo warriors. In 1880, he was commissioned an officer and rose quickly through the ranks. When Díaz needed a band of cutthroats to control his northern border, Kosterlitzky was put in command of the *rurales*. Composed of "reformed" bandits and murderers, the *rurales* and their commander were the law of the North. The Cossack reported only to Díaz whose standing orders were "Catch in the act; kill on the spot."

During 1909, the anarchist newspaper *Regeneración* was urging peasants and workers to rise up and defeat Díaz. Hundreds of copies were distributed clandestinely by smuggling them on board the Mexican railway system and then dropping them off at remote locations. They were then picked up by revolutionary agents and funneled to the Mexican public. Holmdahl probably was commissioned by Kosterlitzky to report these illegal operations from his vantage point on the railroad. He was playing a double game from the start, ostensibly spying for Kosterlitzky, while continuing to work for the regime's revolutionary enemies. Double and triple agents were not a rare commodity during the Mexican revolution.

In October 1910, Francisco Madero, a competing revolutionary and a well-meaning, but ineffective idealist, raised the banner of revolt against the Porfirian regime. Mexico exploded from the Baja in the north to Morelos in the south. Soon ill-trained, but brave recruits, marched to rebel encampments singing:

> *"Mucho trabajo,* (Much work
> *Nunca dinero,* Never any money
> *No hay frijoles,* No beans
> *Que viva Madero."* [13] Long live Madero.)

Because of the young ex-sergeant's military experience, Holmdahl was given the job of guarding the railroad's gold shipments. Since Mexican paper currency was distrusted by almost everyone, workers on the line and, more importantly, landowners selling rights-of-way, insisted on payment in gold bullion or coin. Holmdahl recruited a "brigade" of 200 mounted men who, while escorting the railroad's gold wagons through bandit-infested country, fought off numerous attempts to "liberate" the shipments. Using the same ruthless tactics he had learned in the islands, he hunted down *bandidos* and "never left a live enemy." Soon the route of the railroad track was littered with bodies left to rot in the desert, and the Holmdahl legend was born.

One night in late October 1910, his camp was raided and more than a hundred horses were stolen. Mounting a large posse, Holmdahl tracked the stolen horses across the desert, finally managing to surround the herd and its inept wranglers. The horse thieves didn't put up a fight and surrendered. These men, Holmdahl realized, were not typical *bandidos*. When the men were brought before him, Holmdahl looked at them in the thin cotton pants and shirts of the peasantry and demanded, "Why did you fools steal my horses and why should I not hang you"?

In an excited gabble, the peons explained that Francisco Madero had started a revolution to liberate them, and they wanted the horses, not for themselves, but for the cause of liberty. To the surprise of his brigade, who were already tying hangmen's knots and searching for suitable trees, Holmdahl not only listened to the tales of injustice and terror perpetrated under the dictatorship of Porfirio Díaz, but he actually sympathized with the men. "Not only will I pardon you, I will join you," he told them. Returning to the railroad offices, he resigned his job and took his pay in horses. Now openly a revolutionary, he recruited a motley band of peons to fight against his former boss, Kosterlitzky, and Díaz.

Underneath the hard visage of the professional soldier, perhaps there was a faint ember of the idealist that was touched by the tales of the ragged revolutionists. Typical was the story of Encarnación Acosta, who in later years related

> *I joined the Revolution on November 20, 1910, when I was only thirteen years old . . . I joined more by . . . outrage, and revenge than patriotism. The landlord . . . would often taunt and debase my father . . . The unfair landlord hit my father . . . and hit me too. . . Officials under Díaz. . . took our ranch along with the newborn harvest.* [14]

Nineteen-year-old Francisco Zamora Arce complained,

> *The English and North Americans were the Owners and administrators of the railroads, mining camps, oil, sea, fruit and lumber products. The Frenchmen controlled the clothing industry and the Spanish oversaw the marketed goods. The rich and powerful Mexicans owned the . . . land. We, the poor citizens, owned nothing.* [15]

From the impoverished villages enraged peasants shouted "Venga a la presa" ("come join the fight") and "Muerte a Porfirio" ("death to Porfirio"). Some who knew the exciting young Americano yelled, "Vámonos a Holmdahl" ("Let's go with Holmdahl") and the revolution was on.

In early 1911, his rag-tag band attacked and captured a number of West Coast villages held by small garrisons of federal troops.[16] Whether his men were loyal to the Flores Magón anarchists or to the Madero liberals is not known. Probably, at that time, neither Holmdahl or his men gave a damn. They were fighting the federals and the *ricos* and suddenly there was hope and a chance for glory.

After instigating a failed jailbreak in his old stomping grounds of Tepic, which resulted in the execution of more than 300 prisoners by the Díaz forces, Holmdahl again fled to the hills. A few weeks later with a band of twenty-two men, he raided the *Buena Noche* mine near Rosario and made off with twenty-seven cases of dynamite, with which he started a bomb factory at his mountain hideout.

After a sufficient number of bombs were constructed, Holmdahl was ready to attack Rosario. It was during this time that Holmdahl joined the forces of Martín Espinosa, who was elected "general" by the swarms of peasants flocking to join the revolt. Espinosa's men quickly captured Rosario and then moved toward Rosamorada, where Holmdahl earlier had been imprisoned.

By now their army numbered more than 3,000 men; many were bandits and some were armed only with machetes. After a few days of hard fighting they took the town, but Holmdahl was disappointed to find that the priest who had visited his cell had fled in terror to Tepic. After the customary executions of captured troops some semblance of order was restored. That night Holmdahl wrote that the revolutionary army got roaring drunk on tequila. The bandit element, he said, decided to liberate the almost 700 prisoners held in the city jail.

Since only a few inmates were political prisoners and the rest were murderers, rapists, and thieves, Holmdahl turned back the mob, telling them the prisoners were to be freed in the morning as soon as new clothing and funds could be accumulated to give them a new start in life. Then he went to General Espinosa with the dilemma—if they released most of the hardened criminals, they would let loose a terror of rapine, murder, and theft; if they didn't release them in the morning, many of their troops might revolt against their leaders and attack the jail.

As Espinosa and his staff pondered, Holmdahl, always the practical man, came up with a solution. "Why not," he said, "look at the prison books, find out who are the worst murderers, take them out at midnight and shoot them. We won't use regular soldiers for the firing squad, we'll use officers." Espinosa agreed. After studying prison records, 112 of the worst killers were selected for execution, and six officers were chosen for the firing squad. In small groups, those selected for shooting were told they were to march to the little town of Acaponeta, where, if they joined the ranks of the revolutionary army, their sentences would be commuted.

The happy thugs were then marched out of the prison with an officer escort. On the road to Acaponeta they passed the local cemetery. There they were halted and promptly shot. "This kept us busy the whole night," Holmdahl wrote.

The next morning more than 500 of the least noxious prisoners were released as the army cheered. They were given new clothes looted from the town and five pesos to start a new and honest life. When it was noticed that not a few prisoners were missing, Espinosa casually remarked they been transferred to an army unit at Acaponeta. One can presume as a result of the previous fighting there were enough unburied bodies at the cemetery so that the slain prisoners attracted no notice. Then again, probably no one gave a damn. "Many of the freed turned out to be fine citizens but others later had to be executed after a military court martial," Holmdahl wrote.

After several months spent in cleaning up coastal towns that were still loyal to Díaz, the revolutionaries entered Tepic, and General Espinosa ensconced himself and a growing entourage in the Governor's Palace. As things quieted down, Holmdahl became aware that Espinosa was beginning to plot against Madero's army. One night, he and seven other officers were brought before the general who asked them to join his *junta*. They refused and that night they wisely fled to the mountains, where they joined a band of 280 Cora Indians loyal to Madero. Shortly thereafter, Holmdahl, the seven officers, and the Indians, armed with bows and arrows and an old brass cannon, attacked Tepic. They had presumed that Espinosa's men would defect and join their cause. They didn't.

After thirty-six hours of hard fighting, the Madero loyalists were soundly beaten, and Holmdahl and his compadres again fled to the mountains "leaving more than two-thirds of our men lying dead on the streets of the city."[17] It was probably during this fight that Holmdahl was wounded by a shell that burst near him, killing the man standing next to him. The wound was not serious and within a few weeks he was back at the forefront of the fighting.

By this time, the decisive battle of the revolution against Díaz was building in the northeast, where Madero had his fledgling army camped near Casas Grandes, Chihuahua. In the spring of 1911, Holmdahl joined Madero's army of peasants, former bandits and a smattering of American volunteers near Juárez, just across the border from El Paso, Texas.

In May 1911, Madero troops led by Pascual Orozco, a Sonoran mule-skinner, with additional help from bandits commanded by Pancho Villa and a brigade of American volunteers, attacked the Federal stronghold at Juárez. By this time, Holmdahl, whether from ideology, a sense of adventure, or for cold, hard cash, was committed to Madero's cause.

After a hard fought, three-day battle, the Federal forces surrendered, breaking the back of the regime and sending Porfirio Díaz into permanent exile. As the old dictator boarded the ship taking him to Spain, he remarked, "Madero has unleashed tigers. Let us see if he can control them."

Following the surrender of Juárez, a biographical sketch obtained from the Mexican Consulate in El Paso, Texas, states that Holmdahl was named a "Captain of the rural garrison of that city." The sketch further states that during 1911, probably in the latter part of May and June, he fought alongside troops loyal to Madero in the states of Sonora, Sinaloa, Jalisco, and Tepic on Mexico's western coast. These were military operations against the Flores Magón anarchists who had rebelled against Madero's forces.[18] It all became a little confusing to the outsider and American journalists often remarked, "You needed a scorecard to keep the teams and players straight." They might have added you needed an update regularly as the players changed sides with dizzying speed.

Yaqui troops armed with bows and arrows and machetes at first fought for Madero; they later turned against him. — Aultman Collection

6

Into Yaqui Country

*Free the river and
drive out the whites.*
— Yaqui battle cry

Back in Juárez, Holmdahl met General Benjamin Viljoen, who was to be his commanding officer in his next adventure. Viljoen was a handsome, soft-spoken man, of above-average height, muscular in appearance, with cool blue eyes and a heavy brown mustache. He was born in South Africa, the son of proud Boer pioneers. He first worked as a policeman and later became an editor of an anti-British, pro-nationalist newspaper in Johannesburg.[1]

When the British and the Boers came to blows in South Africa in 1899, Viljoen, then thirty-two years old, turned into a defiant hotspur urging his countrymen to battle. In a memorable article he wrote, it was time to "Put trust in your God and your Mauser." Many did, and for two years Viljoen led Boer Commandos in vicious fighting against the forces of good Queen Victoria.

At the battle of Vaalkrantz, Viljoen with eighty militiamen held off a British force of three thousand regulars for more than seven hours, inflicting heavy casualties. As British numbers began to overwhelm the Boers, he turned to guerrilla warfare and as the head of the Johannesburg Commandos he raided army bases, cut off supply trains, ambushed patrols and generally raised hell with the British invaders.

Soon he was promoted to Assistant Commandant-General for all the Boer forces in the Transvaal. As the war was ending, in January 1902, he was taken prisoner following a British ambush. Along with the most dangerous and recalcitrant enemies of the British, he was transported to St. Helena, Napoleon's desolate island of imprisonment, until the war ended in May.

General Benjamin Viljoen, a former commando leader in the Boer War, planned guerrilla raids for the revolutionaries. Viljoen is second from the right.
— Aultman Collection

Refusing to sign an oath of allegiance to the British conquerors, Viljoen, along with a number of Boers, journeyed to America in 1903. After surveying various sites in the United States and Mexico, Viljoen settled in the Mesilla Valley in the territory of New Mexico, close to the border of the Mexican state of Chihuahua. From a bold and daring guerrilla leader, Viljoen made the difficult transition to a peaceable farmer, cultivating a seven-hundred and fifty-acre tract he called "Hope Harvest" near the small border town of Chamberino.

In 1908, Viljoen became an American citizen and later was named a member of the delegation to Washington D.C. requesting statehood for New Mexico. He was appointed an aide-de-camp to the governor of that state with the rank of colonel and later was commissioned a major in the 1st Infantry Regiment of the New Mexico National Guard.

There still must have been a wild streak of adventure in the 43-year-old farmer. When in November 1910, the Madero revolution broke out, Viljoen left his farm and volunteered as a military advisor to the rebels. Madero's forces were clustered near the U.S. border close to the Chihuahuan city of Casas Grandes and later on the western outskirts of Juárez. Viljoen and a brother-in-arms from the Boer War, Captain Jack Malan, organized fast-moving rebel commando units which harried the forces of the dictator.

In May 1911, General Pascual Orozco overruled a timid Madero and attacked Juárez. After several days of bitter fighting the federal commander, General Juan Navarro, hoisted the white flag. Madero ordered Navarro to surrender the garrison to his two mercenary soldier advisors, General Benjamin J. Viljoen and Colonel Giuseppe Garibaldi, the grandson of the famous liberator of Italy. They made a strange twosome since they hated one another. The thiry-one-year-old Garibaldi had fought in three previous wars before joining Madero. He campaigned with the Greeks against the Turks, joined a Venezuelan rebel army in a revolt against their government, and during the Boer War volunteered in a British regiment fighting against Viljoen's commandos.

Aside from the smoldering resentments forged during the South African campaigns, the two were an unlikely looking pair of soldiers. Viljoen looked bulky in the rough clothes of a Mesilla Valley farmer, while Garibaldi, to say the least, had flair. Wearing a green Alpine hat with a bright feather in the hatband, whipcord riding breeches, a Norfolk jacket and cravat, he looked like he was about to join a British lord's grouse hunt.

To hear Garibaldi tell it, he was the brains of Madero's army and he dismissed Viljoen in his book *A Toast to Rebellion*, stating tartly,

> *A General Viljoen, a Boer from the Transvaal and a veteran of the South African War arrived . . . and declared he was military advisor to the president. In this imaginary capacity he gave out a number of interviews, but we regarded him as a harmless crank.*[2]

In written accounts and verbal interviews about his Mexican service, Viljoen pretended that Garibaldi never existed.

When the Díaz regime fell and the aging dictator fled to Spain, most Mexicans celebrated, thinking the revolution was over. They could not have been more wrong. There was, as usual, a Yaqui problem. In the late summer of 1911, Holmdahl, still a captain of *rurales*,

Giuseppe Garibaldi (second from right, standing), grandson of the founder of the modern state of Italy was a military consultant serving under Madero.
— Aultman Collection

was ordered to accompany General Viljoen to Guaymas, Sonora. Their mission was to try to convince the fierce Yaqui natives of that region to make peace and cooperate with the Madero government.

The Yaquis were an anomaly among Mexican tribes in that they had never bent their heads and accepted the yoke of the Spanish overlords. They were equally rebellious against every Mexican regime. The result was intermittent guerrilla warfare against both Mexico City and the Sonoran governments that had gone on for more than 400 years. By all odds, by the turn of the twentieth century, the tribe should have been exhausted.

Their most recent ordeals began in the 1880s under the Díaz dictatorship. Then rich *hacendados* and politicians fomented plans to steal the rich Yaqui lands in southern Sonora through legal chicanery backed by brute force.[3] As usual the tribesmen resisted. The Yaqui men were a hardy lot, according to John Kenneth Turner, who in his seminal work on pre-revolutionary conditions in Mexico, *Barbarous Mexico* described them as broad-shouldered, deep-chested, with sinewy legs and much taller than average Mexicans.[4]

Rarely armed with modern rifles, the Yaqui warrior went into battle with a very lethal bow, a quiver of arrows, and a machete. Primitive though these weapons were, they were not taken lightly by the Mexican regulars and *rurales* who faced them. The Yaqui's bow was four and a half to five feet in length and their arrows were three-feet long with razor sharp points that could gut a soldier as handily as a slug from a .30-.30. They fired their arrows with a high trajectory, so they came raining down almost vertically, making it almost impossible to spot the position of the bowman. At night and up close, a machete in the hands of a Yaqui was every soldier's nightmare. Sometimes they fought clothed only in a breech-cloth, but often wore shirts and knee-length pants of light cotton. Some had white dots painted on their foreheads, while others wore a colored headband to hold back their shoulder-length black hair.

They attacked day or night to the eerie beat of a shallow wooden drum over which goat hide or sheepskin was stretched. If the bagpipes of the Scottish Regiments of Her Majesty's Britannic armies sent shivers down the spines of enemy troops, the deep thumping of the Yaqui drum echoing through the jungle or down a mountain arroyo created no less a feeling of near panic in their opponents. A British soldier of fortune serving in the Mexican army, I. Thord-Gray, recalled that they chanted war songs as they went into battle and "moved like ghosts as they disappeared into the brush . . . country of Sonora, Sinaloa and Tepic." Thord-Gray continued, "There, the bow was supreme . . . and Federals with modern rifles and machine-guns were often forced to evacuate their positions."[5]

After twenty-five years of warfare with the armies of Díaz, however, the Yaquis had suffered atrocities on a par with those of the Holocaust. In 1892 an army general, frustrated by the Yaqui guerrilla tactics, entered the little town of Navajo, and, according to a Mexican historian, hung so many men, women, and children that he ran out of rope. It was necessary to cut down the dangling corpses and reuse the rope five or six times until the general's ire was satiated.

In another incident, a colonel roped together dozens of men, women, and children, loaded them on the gunboat *El Demócrata* and sailed out of the mouth of the Yaqui River into the Pacific Ocean. At sea, he tied a heavy weight to the end of the rope, and, while the bound Yaquis were clustered on the deck, the weight was tossed

Yaqui troops were frequently accompanied by wives and girl friends during campaigns in Sonora. (1913-1914) — Aultman Collection

overboard. Like a row of dominos, the doomed tribesmen toppled into the shark-infested waters. There were many such reports. [6]

But perhaps they were the lucky ones. By the turn of the century, when the Mexican army became equipped with excellent German Mauser rifles, the tide of battle turned. After a Yaqui army was decimated by mass firepower, thousands of the tribesmen were transported to the Yucatán Peninsula on the southeast coast of Mexico. There, they were worked in slave labor conditions on the henequen plantations until they died of disease, malnutrition, and overwork.

Yaqui lands were seized and many of the remaining tribesmen were scattered throughout Sonora to work as laborers for the local *ricos*. Some, called *broncos*, escaped into the Bacetete Mountains, where they carried on intermittent raids, swooping down on small army or police patrols and exacting a terrible revenge on those not fortunate enough to be killed instantly. In 1904 one of the Yaqui chiefs, Luis Buli, and a band of followers were recruited, or dragooned, into federal service as auxiliaries to the local *rurales*. When the revolution broke out, they half-heartedly fought on the side of Díaz; the majority, however, sided with the forces of Madero.

Hailing the egalitarian rhetoric of Madero and his adherents, in the early days of the revolution Yaquis served the cause faithfully and with great heroism. When Díaz fled and Madero seized the reins of government, they expected the immediate return of all the land stolen from them over the previous decades. This, however, was not to be.

A Yaqui delegation journeyed to Mexico City demanding the return of their land and the immediate freedom and transportation back to Sonora of their brothers still surviving in the hell of the Yucatán plantations. The stalling of the ultra-legalistic functionaries of the Madero regime soon frustrated the tribesmen, while legal double talk and requests for non-existent title deeds and other documents brought them to a fine fury.

Returning to Sonora in disgust, they began seizing their former lands by armed force. Roaming the Yaqui River valleys of their homeland, they raided Mestizo haciendas, shot up towns, and rustled cattle under the cry: "*Río libre y fuera blancos*" ("Free the river and drive out the whites"). Madero, a compassionate man, sympathized with the tribesmen and attempted a reconciliation. Trying to explain the complications arising from title deeds now held by Mexican and foreign investors, he pleaded for time to straighten out complex land-holding problems.

To talk reason with his disgruntled former allies he appointed General Viljoen as "Commissioner to the Yaqui Tribe." Accompanied by Mexican army troops and Capitan Primero Emil Holmdahl, commanding a column of *rurales*, the General trekked to Guaymas on the Pacific coast. Arriving at that tropical seaport, Viljoen had an unsatisfactory meeting with the Yaquis who refused to disarm. After a series of negotiations some land was returned to them and several hundred tribesmen were repatriated from the

Yucatán. But it was not enough, and the tribesmen continued to raid the Mexican and American-owned farms and ranches throughout the area.

Holmdahl with his *rurales*, a few Pima Indian scouts, and regular units of the old Díaz, now the Madero army, patrolled the rugged terrain of the Yaqui Valley in a vain effort to stop the spreading violence. But as one Mexican general complained, "They [the Yaqui fighting men] do just as they please."[7] Under their tough war chiefs, parties of as many as a thousand warriors carried on the fight, hitting the valleys and then running to refuge in the mountains before the frustrated Madero forces could reach them. Viljoen, coming under heavy criticism from other Mexican generals, soon resigned and left the country.

During an interlude in the fighting, Holmdahl had occasion to play Good Samaritan. In February 1912, apparently oblivious to the fact that there was a revolution raging in Mexico, the Cadillac Automobile Company came up with a plan to garner publicity for their new model. The plan called for a 3,000-mile drive from Los Angeles over the rough, almost non-existent Mexican roads all the way to Mexico City. An internationally known race driver, T.J. Beaudit, and a mechanic were assigned to make the trek.

They drove to San Diego, crossed into Baja California, then started down the west coast of Mexico. They managed to cross part of the Sonoran desert and somehow avoided being scalped by the warring Yaquis. In Sinaloa they were nearly shot by marauding rebels and were robbed by *bandidos*. They reached Tepic exhausted. There the mechanic drank some bad water and collapsed with fever. Distraught, Beaudit was about to abandon the project and ship himself and the Cadillac back to Los Angeles. Enter Holmdahl. At the head of his mounted men, Holmdahl rode up to the discouraged race driver, and after Beaudit recounted his tale of woe, Holmdahl secured leave and agreed to accompany the driver as guide and mechanic.

With the proper military passes, the two drove through the rugged country another 500 miles to Mexico City. When the two arrived in the capitol, newspaper photographers' flashbulbs

Into Yaqui Country

recorded the historic event, and both American and Mexican news journals wrote extensively of the daring driver and his soldier guide.

On March 1, a massive luncheon was held for Beaudit and Holmdahl at Mexico City's St. Francis Hotel. The Mexico City English language newspaper *The Daily Mexican* wrote, "The honored guest was E.L. Holmdahl, the young machinist and guide who piloted Mr. Beaudit through the jungles and mountains from Tepic to this city."[8]

T.J. Beaudit (behind the steering wheel) and Emil Holmdahl in front of the Mexican Herald building, 1912. — Holmdahl Papers

Festivities over, Holmdahl returned to Yaqui country to continue the routine business of hunting and killing Indians. He was probably happy when his command was ordered south to quell a more serious revolt. A slightly built Indian, who wore a hat almost as wide as he was tall, had roused the countryside in the state of Morelos into a flaming rebellion against the fumbling Madero government. His name was Emiliano Zapata. For Holmdahl it would be a trip from the frying pan to the fire.

Emil Holmdahl and his pet dog during the campaign against Zapata. Later, a stray bullet shot the dog out of the saddle. — Holmdahl Papers

⁂ 7 ⁂
The Attila of the South

The only government I recognize is my pistols.
— Emiliano Zapata

In early 1912, promoted to major, Holmdahl was put in charge of 1,000 irregular horsemen under the command of General Juvencio Robles. The troops entrained to Juárez, then to Mexico City, and on the long railway passage there was much time for contemplation by the now-hardened mercenary. The revolt of the Indians of Sonora and Morales led to early disillusionment by many of the naive, if enthusiastic, intellectuals who devoutly believed in the revolution.

Although hardly an intellectual, Holmdahl was nothing if not a realist. The expedition against Zapata assembled troops and supplies in Mexico City in preparation for the march south. With considerable prescience, Holmdahl smuggled the first in a series of letters out of Mexico to his mother in Oakland. She in turn forwarded them to the Adjutant General of the U.S. Army in Washington D.C.

They were written in his fine, flowing hand, on stationary of the Hotel St. Francis in Mexico City and dated March 4, 1912. His first missive was a request for a commission as an officer in a regiment of U.S. Volunteers "in case you should see fit to organize such troops for service in Mexico." Detailing his military record in the U.S. Army and his campaigns in Mexico, Holmdahl wrote, "Speak and read Spanish, know almost every trail from boundary line down, know the way of the people and all about the troops way of fighting." Requesting service in a mounted unit as a scout or guide, he added, "If no commission open will be only well pleased to serve my country in any capacity you may see fit. I am 28 years of age, single and in excellent health."[1]

Six months later, on August 23, the war office replied. Writing to a post office box in Nogales, Arizona, they informed him his letter "has been placed on file . . . for consideration in the event your services should be required."[2] Four years later, they were—desperately. By then Emil had learned what it was like to fight the fanatical forces of Emiliano Zapata, called, not without justification, the "Attila of the South."

Zapata's home in Morelos was in a mountainous country crisscrossed by fertile valleys and precious small streams. The population was ninety percent pureblood Indians, but almost all the fertile land was in the hands of a few dozen Mestizo and Spanish *hacendados*. It was the old story of using legal trickery, forged documents, and hired *pistoleros* to seize the Indians' land and reduce them to peonage. The rich hacienda owners had been seizing land and water from the Indians ever since Cortés conquered the country, but by the beginning of the twentieth century the situation was critical. The very existence of many small villages and land holdings was in jeopardy of total extinction, and desperation was conquering fear of the *hacendado* gunmen.

While the land was rich, the people were poor, although Morelos was the third largest sugar-producing region in the world. But while most of the population lived an economic life of bare subsistance, the prosperity of the land was reflected in the homes of the *hacendados*. They lived with their magnificent imported furniture, their sumptuous, European-style interior decoration, their multi-hectare gardens, their stables for polo and race-horses, and their kennels for hunting dogs.[3]

As a youth, some said, Zapata had been forced into the federal army. If so, it was a serious mistake by the government, because the young Indian absorbed all the military experience he could gain, as well as a knowledge of the strengths and weaknesses of the federal forces. Sinewy, with a face the color of chocolate, Zapata was thirty-one years old when the revolution broke out. A small, independent, landowner, he was chosen to lead local forces against Díaz and later was elected spokesman for his village of Anenecuilco.

Meeting with the victorious Madero in Mexico City, Zapata demanded the lands in Morelos be immediately returned to their rightful owners. Although Madero was sincere and the soul of graciousness, Zapata heard the same legal language that had so infuriated the Yaquis. The stalling and equivocations of the Madero regime had the same effect on the hot-blooded Zapata. Well meaning, Madero "wanted to govern, in order to establish in the republic undergoing convulsive spasms a new government, not a new order." He seemed to want his movement to correspond to the impulse of a nineteenth-century political rebellion, and not the first thrusts of a twentieth century social revolution.[4]

Leaving Mexico City in frustration and rage, Zapata returned to his homeland and proclaimed Madero "unfit to realize the promises of the revolution . . . he is a traitor to his principles." On November 28, 1911, Zapata proclaimed his Plan of Ayala which cried out for "Liberty, Justice and Law." We fight, he said, "so that the people will have lands, forests and water." In a bitter denunciation, Zapata exploded, "Madero has betrayed me as well as my army, the people of Morelos, and the whole nation . . . nobody trusts him any longer because he has violated all his promises. He's the most fickle, vacillating man I've ever known . . ."[5]

Soon the cries which commanded "Long live Madero" and "Death to Díaz" were turned into "Down with the haciendas. Long live the peoples' villages."[6] With this, all of Morelos broke into revolt. The self-proclaimed "Death Legion" of the Zapatistas rode under a banner depicting "Our Lady of Guadalupe" surmounting a coal black, grinning skull and crossed bones. It was a symbol that sent shudders down the spines of the landlord classes from Mexico City to the Pacific Ocean. For eight years Zapata rode at the head of his hordes of horsemen, and God help any federal soldier, spy, overseer or *hacendado* that fell into his hands, for no mercy was given or expected. To the gentry in Mexico City, Zapata was a murdering bandit; to the dispossessed all over Mexico, he became a God.

His hemp-smoking Indian hordes included Guerreros, Otomis, Tlahuicas, Mixtecas, and Zapotecas who had cruel ways of disposing of their enemies. A case in point was the giant "century plant" or maguey, native to the region, which had a knife-sharp stalk which, at the plant's maturity, would grow a foot overnight. A screaming victim was stripped naked and tied belly down over the stalk. During the night the hard-tipped point of the stalk sprouted upward, driving itself through the stomach until, by morning, it was projecting through the victim's back.

By dawn's light the eerie spectacle of an impaled corpse, mingled with the blossoming flower tips of the giant plant, greeted the early riser. If there were red ants in the vicinity, only a skeleton remained.[7] As a column of Federal troops approached Zapatista-controlled land in south-central Mexico along its poor roads and worn trails, they were often greeted by the mangled corpses of their recently slaughtered predecessors. Unlike other parts of Mexico there was no cheery music in the land of Zapata.

Journalist and secret agent, H.H. Dunn, in the memoirs of his service with Zapata entitled *The Crimson Jester*, wrote, "Some of the dead men were tied to sharp-needled cactus plants, some crucified on trees, some staked out over the nests of huge red ants . . .and some sewed up in strips of wet rawhide and left to dry in the sunshine."[8]

As Zapata's rampaging bands swept over Morelos and adjoining states, huge landed estates were handed over to the peasants with no hearings, no legalisms, no braying lawyers—the land was given to the people. They, in turn, lived for Emiliano Zapata and would gladly die for him.

The Zapatista tactics were simple; surprise was the key. Concentrating quickly, they made a "gran golpe" of a cavalry charge with wave after wave of horsemen riding at full gallop. The attack was so swift that the federales were usually unable to bring their artillery and machine guns into action before they were overrun. Some daredevil riders drove into enemy lines so fast they lassoed machine guns and artillery limbers and rode off with them before the befuddled gunners could fire.

They were bold and innovative, and as Zapata proudly recalled in later years, the revolution always armed itself with guns and bullets captured from the *haciendas* and the federal army.[9] Before attacking one city, Zapata spotted a defensive line of federales stationed along an aqueduct that ran through the center of town. Somewhere he found a large supply of gasoline, dumped it into the aqueduct stream and as the gasoline flowed by the enemy soldiers, lighted it. A huge sheet of flame cooked the enemy troops while his men, unscathed, swept through and captured the city.

When the Zapatistas took a city—and they hated cities—they emptied the jails, rode to the courthouse, and ceremoniously burned all local records of debts, imprisonments, and false land claims. They appropriated food, horses, money, and arms, drank all the liquor in town, shot all their enemies, and on the way out, ripped out the telephone and telegraph lines. They did not attempt to hold cities; they ambushed and retreated, blocking railways and burning crops. It was classic guerrilla tactics based on the support of the entire rural population.[10]

This was the cauldron into which Holmdahl and his troopers were poured. Not one to give casual compliments, Holmdahl wrote,

> *I was ordered out in the field against the toughest man in Mexico. General Emiliano Zapata is one of the shrewdest men in the Republic and one who does not know fear. I put in some of the hardest service that I have ever experienced in my life [in Zapata country].*[11]

On April 21, 1912, Holmdahl wrote his second letter from Torreón. In it he stated that rumors of American intervention in Mexico were making "it dangerous for every American in this country." "I have no kick coming," he wrote, "I went into this service fully realizing what chances I was taking."[12]

Again recounting his military campaigns in Mexico and offering his services to the United States, Holmdahl warned,

> *The Mexican government is enlisting a great many Japanese military men into the ranks of soldiers. I have seven in my troop and they are all graduates from military college . . . also one who served as an officer in*

Japanese-Russian War. These men are far too intelligent to work for $1.50 Mex per day as a common Mexican soldier.

He wrote he would keep "a good eye" on the Japanese because "should the U.S. start to come in (to Mexico) they would have to fight them."[13]

Holmdahl reported he expected to take part in a battle on April 30, after which "I will march with a machine-gun detachment and 100 men to the states of Sinaloa and Tepic to reinforce the Federales. The more I can kill the less the U.S. will have to take care of." In conclusion, Holmdahl wrote, "My position is very risky so destroy this letter . . . anytime I can be of service to my country please call—if I do not get killed." He signed it "1st Capitan Mexican *rurales*."[14]

In his letter Holmdahl might have used the "yellow peril" threat to enhance his standing as an important observer with the U.S. War Department. After the "Gentlemen's Agreement" of 1907, in which the United States and Japan agreed to prohibit Japanese immigration to the U.S. mainland, relations between the two countries began to deteriorate. There were rumors that thousands of Japanese were immigrating to Mexico, and that they might form a fighting force against American troops if war broke out between Japan and the United States.

In 1911 there were reports of 50,000 Japanese carrying on military maneuvers on the Mexican west coast. It was, of course, nonsense and most of it originated from the fanciful designs of the German foreign office. Their efforts to foment trouble between the United States and Mexico five years later became a major cause of the United States' entry into World War I against Germany. Holmdahl, however, had more immediate concerns than strategic speculations about the Japanese.

On one occasion, Holmdahl was ordered to Cuernavaca, the capitol of Morelos, in the heart of Zapatista country, to bring out a trainload of women and children "who were being abused by the Zapatistas."[15] With an escort of twenty-seven *rurales*, he reached Cuernavaca, loaded the terrified women and children on a train, and headed for Mexico City. It was not to be an uneventful journey. As the train slowed around a curve at Parque, Holmdahl spotted a swarm of Zapatistas attempting to block the track.

Ordering the engineer to stop the train, Holmdahl and his men leaped to the ground and opened a deadly rifle fire on more than 300 Zapatistas who took up firing positions around the track. Alerted by a yell from his lieutenant, Holmdahl turned to see the train's engineer had panicked, and raced full-speed through the Zapatistas going hell-bent for Mexico City and safety.[16]

It was a mixed blessing. The women and children were safe, but Holmdahl and his score of men were abandoned to their fate. He later wrote, "It seemed like certain death as we were outnumbered 11 to 1."[17] As the train raced away, the furious Zapatistas turned on the small band of *rurales*. Holmdahl recounted, "They mounted their horses, let out a yell and made as pretty a cavalry charge as you would wish to see. We met them with rapid fire from our Mauser carbines and checked them." He wrote his *rurales* were "deadly shots" and would fight to the death "as there was no quarter asked or given on either side."[18]

While the besieging Zapatistas had piles of rocks and irregular ground which gave them good cover, the *rurales* had only steel rails to hide behind. Even these few inches of rail were virtually useless as the high powered rifle bullets could penetrate the thin upright part of the rails.

Soon the *rurales'* fire slackened as they took increasing casualties. Under cover of volleys of rifle fire, the Zapatistas began to move forward in short dashes until they got within hand grenade range. The *rurales* were showered with homemade grenades constructed from tin cans filled with explosives. This concoction was put into a rawhide pouch filled with nails, screws, rocks, or whatever was handy, a fuse was stuck into the explosives, and the whole devil's brew was ignited by a Zapatista cigar.

Holmdahl wrote,

> *I was lying on my stomach and hugging the ground as close as I could, when a grenade landed on my arm, next to my face. I couldn't pull the fuse as it had sunk into the hide. I tried to throw it, but as I was lying flat I couldn't throw it very far. Then there was an explosion. It seemed like the*

world came to an end. I was blinded for a moment. There was a terrible pain in my left side.[19]

Later, while recovering in a hospital, he learned he had suffered two broken ribs, both hands were badly burned, while sand and cinders had been blown into his face and arms. At the time, however, he didn't have the luxury of patching up his injuries, as the Zapatistas launched another charge which was barely beaten off. For several hours more the beleaguered *rurales* fought back repeated attacks, but as it began to get dark, Holmdahl realized the next attack would probably annihilate his small band.

But his luck held. Just as darkness fell, he heard the hooting of a steam whistle, and roaring down the track came a train loaded with federal cavalry. They had been dispatched from Mexico City after the panic-stricken refugee train arrived and the crews told of the *rurales* abandonment. As the cavalry-loaded train came to a grinding stop, boxcar doors swung open, and mounted troopers under the command of a Colonel Peña rode out at the charge. The Zapatistas quickly mounted their horses and rode breakneck for safety toward the surrounding mountains, while the *rurales* cheered and a bleeding Holmdahl realized he would live to fight again.

It was morning before the relief train returned the survivors of Holmdahl's little band to Mexico City. There he was taken to a hospital and finally received medical attention for his multiple wounds. It was, he wrote, more than three weeks before he was released from the hospital to take the field again.[20]

By the summer of 1912, Zapata's legionnaires controlled most of southern Mexico and had defeated every federal force sent against them. He boasted, "The only government I recognize is my pistols."[21] In desperation, Madero selected one of the most despised men in Mexico, General Victoriano Huerta, to initiate a new campaign against "The Horde." A full-blooded Indian, Huerta was as ruthless as any Zapatista.

With a shaven head, a drooping mustache, and squinty-eyes peering from behind thick, tinted eyeglasses, the squat general looked somewhat like the fictional ogre Fu Manchu. He was as mean as he looked. Huerta drank more than a quart of French cognac a day, from the first slug upon awakening until he tossed away a drained bottle in the evening. When he was at least half-sober, he was the best fighting general in Mexico, and he was never defeated in the field. In him Zapata had met his match.

With harsh discipline and fierce courage, Huerta's troops smashed the armies of Zapata in open battle, forcing them back into their previous guerrilla tactics of hit-and-run raids. Fighting under the immediate command of General Juvencio Robles, Holmdahl witnessed, and probably took part in, atrocities that made the scorched earth policies in the Philippines look mild.

Huerta instructed Robles, "The best way to handle Zapatistas is with 18 cents worth of rope to hang them with." Robles smartly responded, "My general, I shall hang them to the trees like earrings."[22] And he did. During the remaining months of 1912, General Robles "sowed terror in the state and introduced a scorched earth policy."[23]

If Zapata was the "Attila of the South," Robles and his troops behaved like an invading Mongol horde. Home villages of Zapatista leaders were burnt to the ground, sometimes with their screaming occupants tied inside the flaming houses. Suspects were shot or hung without trial; crops were burned; and God help those who might have information about Zapata's whereabouts, for they were tortured without mercy.

While dazed, fear-crazed, and starving survivors fled to the mountains, Zapata continued his hit-and-run tactics and often gave as good as he got. But gradually his battered forces were worn down, and he and his men withdrew to their mountain redoubts to rest and regroup. But only for a while.

It was while serving in the south that Holmdahl befriended and adopted a small brown-and-white mongrel dog. During the endless dangerous patrols through rebel country, the little mutt provided a measure of company and amusement to the hard-bitten *rurales* under Holmdahl's command. When on the move, the dog nestled itself comfortably in the saddle between the big saddle horn and his master's lean body. Holmdahl remarked that the pup could maintain its seat even during a gallop over broken terrain.[24]

One morning, Holmdahl and his troop were patrolling near a Zapatista stronghold in the hills around Cuernavaca, when they surprised a small detachment of horsemen. Their massive sombreros and slung rifles identified them as Zapatistas and Holmdahl barked *"Adelante compañeros!"* ("Let's go comrades!"). His bugler blew the charge, his men shouted their battle cries, and deploying, spurred into a wild gallop. The Zapatistas, turned their horses and scrambled into a headlong retreat. During a running gun battle, Holmdahl's men accurately firing their six-shooters began to empty saddles.

The Zapatistas were at a disadvantage, for unlike Hollywood Westerns, twisting in the saddle of a racing horse to fire over your shoulder at a moving target is ineffectual at best. In their favor, however, was a knowledge of every trail and twist in the terrain and soon the survivors began to outdistance their pursuers. In the melee, however, one rebel bullet found its mark, striking the little dog and blowing it off the saddle, killing it instantly.[25]

Holmdahl identified one of the fleeing riders as Zapata himself. Dressed in black *charro* clothes, riding a big, white stallion and turning in the saddle, Zapata was firing his revolver at Holmdahl. The bullets whizzed near Holmdahl's ear as he, in turn, blazed away. One of his bullets hit Zapata "in the arm, near the shoulder, causing him to drop his pistol."[26] An excellent horseman, Zapata managed to stay in his saddle and, galloping furiously, finally eluded Holmdahl's men.

After their bugler blew recall, the exhilarated *rurales* and their exhausted horses regrouped and retraced the route of their pursuit. Along the trail, Holmdahl spotted the fallen revolver, and leaning from the saddle he snatched the weapon from the dusty ground. The pistol, now in the possession of Gordon Holmdahl, Emil's nephew, is a "Russian" Model Smith & Wesson .44 caliber, single action, top-break action revolver, which fired a powerful 246-grain lead slug.

Perhaps contemplating his mortality as he recalled those huge bullets whizzing by his ear, Holmdahl examined the weapon closely. Carved ivory handles replaced the standard-issue grips. On one side was a raised sculpture of the Mexican eagle grasping a snake in its beak and on the other side, scratched in the ivory was,

"EMILIO [*sic.*] ZAPATA GENERAL EN JEFE CUARVACA [*sic.*] MORALES MEX MARZO 4 1911."

Along the top of the pistol grip was a line of thirty-two notches scratched deeply into the ivory.[27]

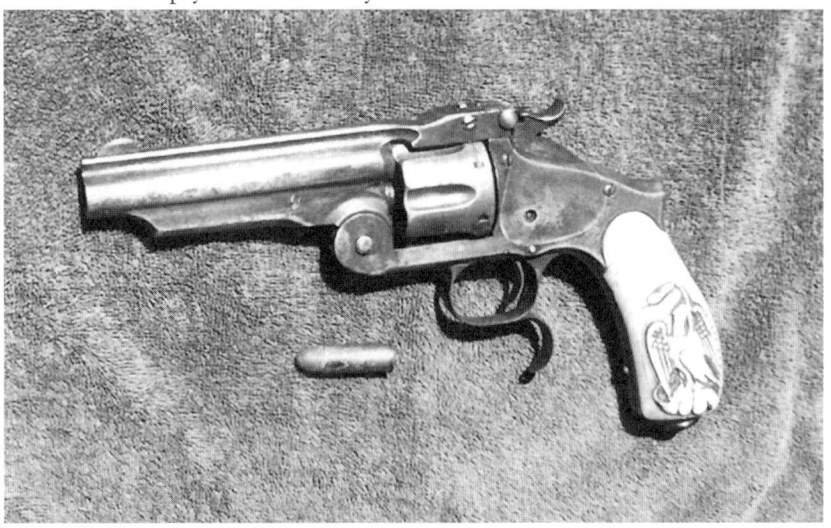

The pistol of Emiliano Zapata captured by Emil Holmdahl. Zapata's name is scratched on the ivory handle. — Gordon Holmdahl Collection

Shortly afterward, the temporary federal successes in the South freed Holmdahl, whose expertise in handling machine-guns and artillery were even more badly needed in the North.

Irregular troops under Pascual Orozco played a major role in defeating the federal garrison at Juárez. — Aultman Collection

❧ 8 ❦
The Orozco Rebellion

Here are your wrappings. Send me more tamales.
— General Pascual Orozco

In March 1912, a serious revolt against Madero's government broke out when Pascual Orozco, the tough mule skinner from Chihuahua who had been the leading guerrilla and fighting chief in the battles against Díaz, threw down the gauntlet. A heavy-shouldered six-footer with a saturnine scowl, Orozco once ambushed a Díaz troop train in Cañon de Mal Paso. After killing the federal soldiers to a man, Orozco had them stripped, bundled up their uniforms, and sent them to Díaz with a note, "Here are your wrappings. Send me some more tamales."[1] This was a man to be reckoned with. It was he who, disobeying Madero's orders, attacked and captured Juárez, and brought about the collapse of the Díaz dictatorship.

It was then, according to Colonel Francisco Gallegos, that Madero made a fatal error. Gallegos, who fought under Madero and later under Villa, in later years, wrote, "When Madero took power he made the mistake of dismissing the majority of the revolutionary leaders and leaving in their same political position many of Díaz's former officials."[2] Among those most alienated was Orozco for after victory was complete, Madero reinstated most of the Díaz officers back into the regular army. He "rewarded the vital military contribution of Orozco with the post of commander of the rural guards of Chihuahua," a post that paid a salary of eight pesos per day.[3]

Holding the dwarfish Madero in contempt and perhaps bought off by money from Luis Terrazas, the richest man in Chihuahua, Orozco claimed the little reformer betrayed the revolution. He launched proposed reforms under his *Plan de Empacadora*. It was one of the many political "Plans" offered by revolutionaries, promising heaven and usually delivering nothing except more hell for the suffering Mexican people.

Orozco endorsed "political and economic reform, including a shorter workday, better working conditions, employment of Mexicans on the railroads, and the return to the villages of lands illegally seized."[4] To many his financing by Terrazas money and promulgating liberal reforms was paradoxical if not deceptive.

It has been rumored that Pancho Villa, who gave much lip service to his "love for little Madero," offered to join the rebellion. But the harsh-tongued Orozco rebuffed him saying, "No common bandits will be accepted into this movement."[5] Another version, however, has it that Orozco urged Villa to join his *junta* but the former bandit refused and pledged his loyalty to Madero.[6] Villa then recaptured his temporarily lost love for Madero and, along with other former revolutionaries, took the field against Orozco's "Colorados" or "Red Flaggers." They were named from the reddish flags they carried, often emblazoned with the slogan "pan y tierra"—bread and land.

Most of the battles between the two forces were fought over control of the Mexican railway system. Practically all of the fighting in the central plateau of northern Mexico was confined to a zone twenty miles wide with the railroad at the center. There were two reasons for the railroad's strategic importance; first, travel over the northern desert country of Chihuahua and Sonora was difficult for cavalry and impossible for infantry. Secondly, the railways transported petroleum, cotton, and copper ore exported to American markets, which in turn supplied vital foreign currency needed to buy military supplies.

Tactically, the contending armies destroyed the railway tracks and bridges as they retreated and rebuilt the road as they advanced.

The Orozco Rebellion

Gregory Mason, a war correspondent during 1914, described the procedure,

> *The usual method is to rip up the rails with a strong iron hoop passed under both of them and attached to an engine by a heavy chain. The engine backs, the rails hold an instant, then come up with a groaning of twisted steel and a rattling shower of spikes sounding like a boiler factory, the anvil chorus, and a dozen machine guns in simultaneous operation.*
>
> *Frequently an engine will tear up an eighth of a mile of rails at one rush before it is obliged to stop by a broken hoop or a snapped chain. When the rails, twisted like grotesque corkscrews, have been accounted for in this fashion, the ties are piled and burned. Dynamite does for the bridges.*[7]

At first Orozco's revolution prospered. He captured Juárez, routed Pancho Villa's troops outside Chihuahua City, and won additional battles at Santa Rosalía and Jiménez. A major setback, how-

During the Orozco rebellion, his troops called "Colorados" carried red flags emblazoned with the battle cry, "pan y tierra" ("bread and land"). — Aultman Collection

ever, which was to prove fatal to the Orozco cause, was a policy announced by the United States government banning arms sales to any of the battling Mexican factions. Since the "Colorados" controlled only the borderlands of Chihuahua, cutting off the flow of arms from Texas and New Mexico did them grievous harm. The federals, however, imported massive supplies of guns and ammunition from Europe through the Mexican ports of Tampico and Vera Cruz.

As Orozco's army became starved of supplies, troops loyal to Madero, commanded by Huerta and Villa, grew fat with modern arms. Soon, Villa's well-armed cavalry rode to the front on the Mexican railways, horses in the boxcars, men on the roof, singing their theme song "La Cucaracha" ("The Cockroach").

> *Oh the cockroach,*
> *Oh the cockroach,*
> *Will not move, the old slow poke;*
> *Because he hasn't,*
> *Because he hasn't,*
> *Any marijuana to smoke.*

As fighting continued, ideals became corroded with bitterness, and the hatred between Villa and Orozco led both sides to new barbarism. When Villa captured enemy soldiers, he always executed the officers, but gave the common soldiers the chance to save their lives by joining his ranks. Not so with "Colorados" who were all shot on the spot. The "Red Flaggers" responded in kind.

As things began to go badly, Orozco drew down the wrath of the United States, when he allowed one of his generals, Inez Salazar, to execute an American machine-gunner. The American, Thomas Fountain, was captured after a Villa retreat from Hidalgo del Parral. That execution, undoubtedly, gave pause to mercenaries on both sides. The Mexican Revolution, they realized, was not a romantic game—it was a bloody fight to the death.

In March 1912, the Madero government, determined to crush Orozco, moved a large force under the command of General González Salas along the railroad from Mexico City north to the rebel stronghold at Hidalgo del Parral. On March 24, the two forces, each with about 8,000 men, clashed. An effective bombardment by federal artillery drove the rebels back from their positions. Retreating rapidly, they fell back along the railway past Jiménez into a mountainous area north of the small town of Rellano.

Sensing victory, Salas ordered a quick advance along the railroad tracks, ordering three troop trains still jammed full of field guns, infantry, and ammunition to move up to the front lines. There they would disembark, and with a quick attack, smash the enemy defenses. It was not a bad tactic, but he failed to reckon with the mad antics of Sam Dreben and Tracy Richardson. These two buddies of the Banana Wars had been recruited by Orozco, who was paying top dollar for anyone who could operate and maintain the new-fangled machine guns.

The rebel army dug in across the foothills bordering the railroad tracks where they dug rifle pits and erected rock walls to provide cover. Believing they were heavily outnumbered, they awaited the federal attack with considerable apprehension, believing their future looked bleak. From a vantage point in the hills, Dreben and Richardson observed the federal deployment.

As sweating troops started manhandling field guns and ammunition from

Sam Dreben, dubbed "The Fighting Jew" by war correspondents, was one of the machine gunners who turned the tide at the battle of Rellano.
— Aultman Collection

the boxcars, Dreben pointed out that the railroad track ran straight through the middle of the federal position. There was nothing blocking the tracks between the two armies. "After they unload those field guns," he said, "they'll shell hell out of us and then the infantry will go right up the middle." Then both men got a wild idea. "Suppose," they pondered, "we get one of our railroad engines, load it up with explosives and drive it right into the Federal trains. It could blow up half their army." They agreed it was just crazy enough to work.

They got an engine, strapped 800 pounds of dynamite to the cowcatcher, and plugged in a dozen detonating caps. With Richardson at the controls and Dreben working like a demon shoveling coal into the boiler, they got up steam and drove down the tracks toward the federal trains. When they got within 100 yards, Richardson said, "I threw the throttle wide open, we leaped out of the cab, and let the engine run wild down the track."[8]

The engine roared into the federal trains with a terrible crash, followed by a series of explosions which set off the ammunition still loaded aboard the trains. Cannon, pieces of boxcars, and parts of soldiers rained down upon the stunned Madero troops. Tracy wrote, "smoke and earth spouted up like a giant geyser." Completely rattled, the surviving federals, dazed, spilled out of the trains, while the troops already in the front line panicked and began to run to the rear.

Slashing with their sabers, federal officers opened fire on their retreating men and shot many in the back before the retreat was stopped. Rallying the infantry, they drove them into line and attacked the entrenched rebels. Dreben and Richardson, meanwhile, dashed back to their own lines and took position with their machine guns. Richardson wrote,

Tracy Richardson fought in a dozen wars and was called the "World's Greatest Machine Gunner."
— Douglas V. Meed Collection

> *Sam Dreben and I . . . could work a crossfire from our machine guns. Those poor Federal soldiers were marched up against us in close formation Rank after rank, Sam and I mowed them down until it sickened us.*[9]

As the federals fell back in confusion, Orozco's cavalry charged, making shambles of the retreat as hundreds of frightened soldiers leaped aboard the still operating third train. The frightened engineer backed the train down the tracks and didn't stop until he deposited the disorderly mob at Hidalgo del Parral.

Madero's broken army abandoned all their artillery, machine guns, hundreds of rifles, and a large store of ammunition that had not blown up. Richardson said later they counted 1,200 dead near the scene of the explosion. "Our losses" he reported, "were only 20 killed, 100 wounded."[10]

The remaining federal train, with troops crowded into boxcars and some hanging perilously from the roofs, fled more than 100 miles south to Torreón. There, General Salas, overcome with grief and shame, put a revolver to his temple and blew out his brains.

President Madero became badly frightened by Orozco's early successes, and in an action he would later regret, he gave the sinister General Huerta overall command of his forces in the North. In the reorganization that followed, in April 1912, Holmdahl was ordered to report to General Gerónimo Treviño in command of the Third Military District, headquartered in Monterrey. There he was assigned as commander of the 5th Regiment of Cavalry. General Treviño, once a confidant and ally of Díaz, was now a rich landowner. In the 1870s he had cooperated with American cavalry forces during campaigns against Apache Indians and Mexican bandits. Later, he married the daughter of the U.S. General E.O.C. Ord. By a curious twist of Mexican politics, he now supported Madero against Orozco.

Holmdahl described Treviño as "one of Mexico's oldest and best generals but too old to take the field." He characterized Orozco as a man who "betrayed every confidence placed in him . . . and one of the biggest four-flushers that the war produced." Holmdahl, for all his cynicism, was probably ideologically devoted to Madero's cause.

Serving under Treviño, Holmdahl later wrote that his "Carbineros were a fine bunch of young men and were anxious to get to the front." For about a month the regiment skirmished with "Red Flaggers" in northern Mexico.[11]

In May 1912, Holmdahl was assigned to the artillery section of Huerta's army in command of a Maxim machine-gun company. It may well have been Fountain's old outfit, and that mercenary's fate might have crossed Holmdahl's mind.[12] The assignment must have been a welcome one, because Maxim guns were the first easily transported, reliable machine guns. The old Civil War-era multi-barreled Gatling guns, a few of which were still used by the insurgents, were heavy, ungainly, and could shoot only as fast as they could be cranked. They often jammed. Most other early machine guns suffered from one or more similar disadvantages.

The inventor Hiram Maxim designed a single barrel, recoil-operated, water-cooled machine gun that was light, fired simply by holding down the trigger, could spray out 650 rounds per minute, and rarely jammed. First used by the British army, it enabled their small forces to overcome native armies three or four times their number.

Throughout the spring and early summer of 1912, in a series of battles, Huerta's well-supplied and effectively-led troops began to wear down Orozco's dwindling forces. On May 23, General Treviño summoned Holmdahl to his headquarters and asked him to volunteer for a dangerous mission.[13] Following the disaster at Rellano, Captain Lorenzo Aguilar, first cousin of President Madero, and two other officers were reported missing after a bitter fight at the small village of Pedriceña.

Madero was worried about the fate of his favorite cousin and Holmdahl's mission was to travel through enemy lines, locate Aguilar dead or alive, and bring him or his body back to Madero-controlled territory. With false papers identifying him as a correspondent for the *Monterrey Daily Mexican-American*, Holmdahl boarded a train bound for the headquarters of his friend, General Aureliano Blanquet, located a few miles south of Torreón.

When Blanquet learned of the mission, he refused to let Holmdahl cross into enemy territory controlled by the Orozco General Emilio Campa. Campa, a former medical student at an American university, hated Americans. A mean drunk and psychopath, Campa had recently trumped up charges against Sam Dreben and Tracy Richardson. He arrested the pair and planned to shoot them, in order to "rid Mexico of all gringos."[14] Probably, he was jealous of their reputation for heroism (as a result of their coup at Rellano). After a hairbreadth escape from jail, the two Americans rode for their lives until they reached Orozco's headquarters and safety.

Blanquet feared that, regardless of Holmdahl's papers, he might become a candidate for Campa's firing squad. There was additional danger in the fact that Campa knew Holmdahl as an old comrade in the campaigns against Díaz. Appreciating Blanquet's concern, Holmdahl, nevertheless, bought a horse and slipped out of camp, riding thirty-five miles north, and reaching Hacienda Refugio. There, unfortunately, he encountered a "Red Flagger" patrol and was taken before General Campa.

Campa greeted him like an old friend, but then questioned him, asking if he were still in the service of Madero. Holmdahl denied this and presented his newspaper correspondent's papers. The mercurial Campa smiled, then called him a liar and a spy, and announced he would shoot him. Again, Holmdahl's ability as a con artist probably saved his life. Talking fast, he "half convinced" Campa of his *bona fides*. Campa smiled and treated Holmdahl to a delicious meal at his officers' mess. In the morning, however, he refused to let the "correspondent" pass through his lines.

Disappointed, but lucky to be alive, Holmdahl left the camp, skirted the rebel patrols, got close to Pedriceña and was picked up by another scouting party. Arrested, he was brought before another rebel general, and again he talked his way past him and finally reached Pedriceña.

There he found an old *rurale*, who recounted a sad story. During the fighting on May 14, Aguilar searching for ammunition for his beleaguered men, ran into "Red Flaggers" disguised in federal uniforms. When he approached them they shouted, "Viva Madero," but when he got up close they leveled their rifles, shouted "Viva Orozco" and took him prisoner. Two other officers and several dozen men were captured when they ran out of ammunition.

Another witness, Señora María Peña, told Holmdahl that on May 15, about 5:30 in the morning, she heard loud voices in a field behind her house. Going outside she saw six federal officers lined up in the field surrounded by "Red Flaggers." She said an Orozco officer told the men if they shouted "Viva Orozco," their lives would be spared, but defiantly, the prisoners shouted, "Viva Madero." They were promptly shot and their bodies dragged to a nearby ditch and dumped in with the rest of the casualties of the battle.

Holmdahl purchased a shovel and a mule, and that night he went to the mass graveyard where his witnesses said Aguilar was buried.

There he started digging and by lantern light examined each body he dug up. The sixteenth corpse he exhumed was the young captain.

Recovering Aguilar's body, he strapped him on the mule, and then he rode ninety miles through enemy lines until they got back to territory under Federal control.[15] He returned the body to Monterrey, and there Holmdahl, General Treviño, the mayor of the city, and other high ranking officials posed for pictures around the flower-draped casket of the unfortunate young captain.[16]

On August 9, 1912, Holmdahl wrote again to the U.S. Army Adjutant General in Washington D.C. This time the letter was mailed from the Montezuma Hotel, in the border city of Nogales, Arizona. From there he reported, "Things are looking worse every day down here." Again offering his services, he pointed out he had commanded 5,000 soldiers of all army branches at the beginning of the revolution:

> *Am thoroughly acquainted with their country, climatic conditions, water holes, mountain trails, their mode of fighting and supply stations and will gladly give you any information you wish, as I believe that am better posted than any other American as I have fought with them for two years . . . I am on my way to report to the general in command of the First Military Zone in Sonora.*

He signed the letter: E.H. Holmdahl, Capitan Primero Caballeria.[17]

By September 1912, Orozco's "Colorado" movement was smashed by Huerta and Villa, and the mule-skinner general fled to the safety of the United States. After harrowing escapes, Tracy Richardson and Sam Dreben also made the border safely. For the rest of the year, Holmdahl and his machine guns did mop-up duty in minor campaigns, fighting under Colonel Guillermo Rubio Navarrete in Chihuahua, General Aureliano Blanquet in Durango and Zacatecas, and with General Gerónimo Treviño in Nuevo León, Coahuila, and Tamaulipas.[18] In Mexico, during 1912, machine gunning was a growth industry.

By October, Holmdahl was once again restless. Perhaps he became bored with machine gunning and longed for a cavalry command with the hardened *rurales* he had led against Zapata. On

October 24, 1912, he received a letter from his sometimes mentor and sometimes foe, Emilio Kosterlitzky, the tough Cossack who commanded all the *rurales* in northern Mexico.

It was in reply to a Holmdahl request for a transfer back to the *rurales*. After complaining about a drunken lieutenant in his command, the old "Iron Fist of Porfirio Díaz," now working for Madero, wrote, "Believe me I deeply regret not to be able to have you with me for the present, but hope an early opportunity to notify you of having a place for you. With warm personal regards..."[19] If Kosterlitzky was sincere, and he probably was, it indicated he no longer had the free hand he once enjoyed under Díaz. Madero probably kept him under a short leash for trust was not overly abundant in revolutionary Mexico. For good reason.

But, if the unrestrained license of a *rurale* was unavailable, Holmdahl must have again turned to espionage to succor his thirst for adventure. In late 1912 he entered the shadowy underworld of El Paso, Texas. El Paso was separated from the Mexican city of Juárez by only a short bridge across the Rio Grande. The twin cities were a crossroads for trade between northern Mexico and the U.S. southwest with railroads linking major markets in both countries. El Paso was also a hub of activity for gunrunners, smugglers, war correspondents, and spies. It was a haven for dissident Mexican politicians on the run from a government that plotted new revolts in the coffee houses and cantinas in its crowded south side called "Chihuahuita" ("Little Chihuahua").[20]

A report in Holmdahl's handwriting among his personal papers written in English indicates that in December 1912, he was working undercover in El Paso.[21] The long report, dated December 28, 1912, states that one Jesús Cesneros [*sic*], the proprietor of a barber shop in the 500 block of South El Paso Street, had a secret back room. It was used, Holmdahl said, as a headquarters for renegade "Red Flaggers" who were smuggling guns and ammunition across the border and plotting another revolt.

In the report, Holmdahl lists the names of a half-dozen former Orozco officers. He describes how they subverted the Madero garrison in Juárez by offering the poorly paid soldiers large sums of money in return for turning over their ammunition to one of their spies. The spy, after accumulating fifty rounds of ammunition, would give it to a young woman Simone Acosta, who would smuggle it across the border under her voluminous skirts. The ammunition was stored in a secret cache under the floor of the barber shop. Then it was smuggled back across the border to the rebel army.

Holmdahl obviously had penetrated the "Red Flag" cabal and had his own spy at their meetings. In his report he gave details about a cattle-rustling scheme, the proceeds of which would go to support the rebels. Topics of discussion among the plotters included movements of troops under the command of General Trucy Aubert, still loyal to Madero. The plotters also discussed ways the soldiers could be subverted into joining the Orozco ranks.

The mastermind of the operation was General Inez Salazar. Salazar, a giant of a man, was the chameleon of the Mexican revolution. It was said he could change sides faster than a lizard could change colors, although during the revolution that was not necessarily a unique characteristic among Mexican generals. Salazar had at various times both served under and fought against Madero, Orozco, Huerta, and Villa, and was as quixotically cruel as he was politically unstable. It was he who further aggravated relations between Orozco and the U.S. government when he decided on a whim to shoot the captured American mercenary, Thomas Fountain. At this time he was still plotting with Orozco, but that would soon change.

Holmdahl's report was probably written for General Aubert, but there is circumstantial evidence that he was also providing information to the U. S. Bureau of Investigation. The bureau was the forerunner of the FBI, and was keeping an eye on illegal activities along the Texas-Mexican border. There is a letter dated November 4, 1913, in the Bancroft Library files from an agent in the Douglas, Arizona, office of the Bureau to an agent presumably in El Paso responding to a request for information about Holmdahl's whereabouts. In it, the Douglas agent reports,

> *I saw Holmdahl in Douglas about Oct. 25 . . . He stated to me that he had been quite seriously wounded . . . he was thin and pale, but was wearing good clothes and appeared to be cheerful . . . I do not believe he was*

suffering for the want of any necessaries . . . If such had been the case I surly [sic.] *would have offered him assistance . . .*[22]

Holmdahl must have been up to his neck in espionage at various times since he first entered Mexico in 1909. There is a cryptic note, dated September 9, 1911, in his papers which states, "Meet me on [undecipherable word] when No. 12 gets to Hermosillo—Opr. Nogales can give you time." The note was signed, H.J. Temple. A later addition to the note, in Holmdahl's handwriting, states, "Temple was general manager SP.RR of Mexico. Shot himself when confronted by U.S. Agents making arrest for selling information to Germans." There is no date on Holmdahl's addition.[23]

In several letters to the U.S. War Department, Holmdahl had given information about conditions in Mexico and offered to be a conduit for further information. He was at least intermittently acting as an agent for the U.S. government. There was no ambiguity in his reporting both to General Aubert and to U.S. officials, since the American government supported Madero as the legitimate head of Mexico and considered Orozco a bandit or rebel, at best. Finally, whatever else Holmdahl was, he was a loyal American who was always ready to put his life on the line for his country.

General Huerta, meanwhile, returned to Mexico City as a hero, and after being appointed commander-in-chief of the Mexican armies, he began to plot against his president. While Zapata on the left felt betrayed, the old Díaz rightwing, led by Huerta, viewed Madero as a usurper "with the common aim of toppling the Mexican president."[24] On February 9, 1913, the "Decena Trágica," the tragic ten days, a phony war was staged in Mexico City between conservative forces and federal troops under Huerta. During the intense fighting, innocent civilians were killed until the farce ended.[25] During that time, U.S. Ambassador Henry Lane Wilson acted as a go-between for the contending forces as Wilson, opposed to Madero, supported the coup led by Huerta.

On the night of February 17, Huerta had Madero arrested on trumped-up charges, and on February 22 had him assassinated.[26] Huerta seized power and Mexico again was held in the grip of a ruthless dictator, but Huerta had unleashed a hornet's nest of opposition. "The Maderistas had no intention of allowing the Huertistas to savor their ill-gotten laurels."[27] Venustiano Carranza, the governor of Coahuila, refused to recognize the Huerta regime. With the backing of Pancho Villa in Chihuahua and Alvaro Obregón, a bean planter in Sonora, Carranza went to war against the murderer of "The Apostle of the Mexican Revolution."[28]

While many of the old revolutionaries took up arms against Huerta, a new ally was his recently defeated foe, Pascual Orozco. The Mexican scorecard was by now almost unfathomable. Pancho Villa, after shedding public tears for his "beloved little president," crossed from his refuge in El Paso and started recruiting a new army.

On December 24, 1913, Holmdahl, writing the adjutant general from El Paso, stated he deserted the federal forces in Juárez on February 18, 1913. His reason for doing so was the assassination of Madero. Holmdahl escaped to Sonora and joined the constitutionalist forces, where he was commissioned a first captain of Artillery.[29]

After learning of Madero's murder, and probably suspecting that his old boss General Aureliano Blanquet was involved, Holmdahl decided to desert and return to the United States. Whatever other inducements may have been offered, Holmdahl was always loyal to Madero, and if a cold-blooded soldier of fortune had any passion for a cause, Madero's selfless passion to free Mexico from dictatorship resonated deeply within him.

On that cold night in February, Holmdahl swam his horse across the Rio Grande and dismounted, but while drying himself an American patrol approached, spotted him, and let out a shout. Not wishing to be hauled in as a border jumper, Holmdahl leaped into the frigid Rio Grande. The river was high, and a swift current carried him downstream, washing him up on the Mexican bank. His luck failed, and a patrol of troops loyal to the rebel General Inez Salazar took him prisoner.

By this time, Holmdahl was well known on both sides of the border, and when he was brought before the general, Salazar laughed and said Holmdahl would be shot in the morning. He was promptly thrown into a local prison, but lady luck had not completely turned against him, and he managed to bribe a guard, escaping in the early morning darkness. He now had cheated a firing squad twice; it was not to be the last time.

Holmdahl gave up thoughts about leaving Mexico; perhaps he wanted a crack at Salazar, who was now allied with Huerta in establishing a dictatorship as evil as that of Díaz. Holmdahl traveled to Hermosillo, Sonora, where he joined the army of General Benjamin J. Hill, in rebellion against Huerta. The Yaquis were rebelling again on the west coast of Sonora, and on General Hill's orders Holmdahl again campaigned against them.

After the Yaquis had been subdued, Holmdahl wrote that he was sent to help put down Huerta loyalists in Sonora and Sinaloa. When a number of Yaqui tribesmen changed sides and became allies, he joined with his old foes and returned to Chihuahua. There he was assigned to the Francisco Villa Brigade under the command of General Juan M. Medina.

Pancho Villa led former bandits into Madero's fight and helped capture Juárez in 1911.
— Aultman Collection

❧ 9 ❧
Riding with Pancho Villa

There were three whores
Sitting in a silla
Saying to each other
Viva Pancho Villa.
— Soldier's song

According to the folklorist J. Frank Dobie in *Apache Gold and Yaqui Silver*, Pancho Villa sent Holmdahl on a mission into the mountains of Chihuahua.[1] According to Dobie, bandits seized a silver mine and Villa wanted them dead, and the silver mine worked for the benefit of "the people." As Dobie tells it, when Holmdahl arrived at the mine, the bandits fled, two of them taking refuge in a rock-walled corral at nearby Rancho Guerachic. Spotting them, Holmdahl drew his six-gun and spurred his horse.

Sailing over the wall, his revolver blazing, Holmdahl and the bandits blasted away at each other until the bandits fell dead. The gunfight was at such close quarters that Holmdahl was powder-burned, but otherwise unwounded. After turning over the mine to men loyal to Villa, he returned to his artillery command prepared to fight under the banner of the famed Division of the North.

Pancho Villa was born in Durango in 1878, the son of a sharecropper. During his early life he gained notoriety as a bandit and a rustler, but when the Madero revolt began he was recruited by the liberal politician Abraham González. González believed that Villa's knowledge of the terrain and his leadership abilities would not only prove of great value to Madero, but that his service in a good cause would rehabilitate him into becoming a useful citizen.[2]

Villa's exploits during the revolution soon became the stuff of legend. To some, "He was a monster who fills all Mexicans with shame and will go down in history like a blot of blood."[3] To others he was a "Mexican Robin Hood," who was "an avenger, and a righter of wrongs."[4] "He was that genius of warfare to whom is owed the triumph of the Revolution."[5]

On March 6, 1913, Villa learned of Madero's murder. Villa, with eight men riding beside him, crossed the Rio Grande from Texas into Chihuahua bent on vengeance. He learned that his friend and mentor, Abraham González, governor of Chihuahua, had been dragged from a train and brutally slain by Huerta's men. From that time on the quality of Villa's mercy was rarely strained. Within weeks his charisma, his bravery under fire, his genius for fighting, and the support of the peasantry enabled him to raise an army that, for a time, would be the strongest in Mexico.[6]

During the summer of 1913, in winning battles against Huerta's armies, Villa reached the peak of his success. His Division of the North had a strength of almost 50,000 tough, disciplined troops, fanatically loyal only to the "Centaur of the North." Holmdahl, at this time, was assigned to command Villa's machine-gun detachments, while the old bandit's artillery was ably commanded by General Felipe Angeles, a graduate of St. Cyr, the French West Point. Villa also had good medical service and occasionally an airplane flown by an American stunt pilot.

While Venustiano Carranza was the nominal leader of the anti-Huerta forces, it was Villa and his men who did most of the fighting, and his armies remained the best in Mexico. In short order they smashed federal forces at Guerrero, Bustillos and Casas Grandes. In late August 1913, Villa assembled his forces before the city of San Andrés. The city was garrisoned by a force twice the size of Villa's army and commanded by the competent General Félix Terrazas.

On August 26 Villa attacked. After fighting all day, he was unable to force his way into the city because of effective fire from the federal artillery. The following day, Holmdahl and his machine guns were brought up to the firing line. After pouring belt after belt of ammunition into the enemy trenches, by late afternoon he succeeded in silencing the infantry fire. Martín Luis Guzmán, a writer on Villa's staff, reported that Villa complained to one of his officers, "Señor Colonel, we cannot advance as long as their artillery keeps bombarding us . . . Silence their cannon."[7]

As darkness descended across the battlefield, Villa ordered one of his patented *golpe terrífico* cavalry charges. Holmdahl turned his machine guns over to a subordinate, mounted his horse, and rode to the front of Villa's massed cavalry. He found himself riding alongside the famed "Dorados," Villa's personal bodyguards and the best mounted troops in Mexico, and he charged with them, literally into the cannons' mouths.

As Villa's bugler sounded the charge, the cavalry spurred its horses and, screaming "Viva Villa," rode into the guns while the federal artillery opened fire with canister. As he rode forward shouting, one hand on the reins and the other on his .45 caliber revolver, Holmdahl's hat was shot off by a shell fragment.

Somehow, Holmdahl found himself leading the charge, and, reaching the enemy artillery, he fired to the left and right at cowering gunners as the "Dorados" followed him. Suddenly he felt a shock as a bullet or a piece of shrapnel ripped across his stomach and he lurched from his horse and fell moaning onto the ground.[8] The charge, however, overran the federal defense line, captured the artillery, and battered the defenders into submission.

In his book, Guzmán gave credit for winning the battle to daring and skillful officers, including one "of English ancestry, a certain Hondall or Jontal."[9] This was, of course, Emil Holmdahl, who was honored by the Mexican government which awarded him the Legion of Honor and made him a honorary colonel in the Mexican army. A contemporary pamphlet described Holmdahl's charge as heroic and bold, noting that his courage should be memorialized in marble and bronze.[10] Holmdahl spent six weeks in a Villista hospital, where doctors stitched up his stomach and he quickly recovered his accustomed vigor.

The fleeing federals abandoned almost 1,000 dead as well as losing more than fifty artillery pieces, 400 Mauser rifles, 20,000 rounds of ammunition and seven railroad trains loaded with food, medical supplies, and uniforms. The glory of the victory was soured, however, by the brutal murders of captured troops. According to one of Villa's wives, Luz Corral, Orozco sympathizers had poisoned her daughter. Villa, as outraged father, cried out for vengeance and turned the prisoners over to his faithful killer, Rudolfo Fierro.

Fierro, who called himself a frugal executioner, lined up more than four hundred helpless prisoners in groups of three. Forcing them to hug each other back-to-front, Fierro then strode down their lines firing one shot from a high-powered pistol into each trio, fatally drilling all three bodies. "Look how much ammunition I saved," he giggled to Villa. Everyone thought it was terribly amusing.

After the federal debacle at San Andrés, Villa quickly captured Torreón, and then used his captured trains to move his forces to Chihuahua City. His old enemy Pascual Orozco, now allied with Huerta, was commanding the garrison there. Villa sent him a demand for surrender, and Orozco replied, "Come and take us, you son-of-a-bitch."

Enraged, Villa launched a series of massive cavalry attacks on the city, but each one was repulsed with bloody losses. In one of the *golpe terríficos*, Holmdahl took a bullet in the leg, but apparently it was not serious, because he had rejoined his unit by the time Villa pulled a clever coup.

Leaving a small force to surround Chihuahua City and keep up desultory firing, Villa secretly loaded the bulk of his army on trains, abandoned the city and the hated Orozco, and sped toward Juárez. At each station along the 500-mile-journey, he sent a phony message to Juárez, reporting the progress of a federal train filled with reinforcements. Then he cut the telegraph wires.

The federal garrison at Juárez bought the ruse, and on November 15 the Trojan horse pulled into the border city without opposition. By the following day, Villa had captured the surprised 300-man garrison, and Fierro shot them all. Villa methodically looted the many banks, gambling halls, whorehouses, and saloons in the city. With a large war chest, he bought fresh supplies of guns and ammunition that had been smuggled across the nearby U.S. border.

Within the week, however, the reinforced federal garrison at Chihuahua City had broken through his thin cordon and was heading up the railway toward Juárez. Villa sent out patrols to wreck the Central Railroad line leading to the city and deployed his men in a line centered at Tierra Blanca, twenty miles south of Juárez. There he occupied high ground overlooking the sandy desert through

Villa's irregular cavalry entering Juárez in 1911 after the surrender of the federal garrison. — Aultman Collection

which the federal army would attack. His men dug in on a low ridge of dunes on each side of the railroad tracks.

The enemy force under Huerta loyalist, General Inez Salazar collided with Villa's troops on November 24 at Tierra Blanca. The battle would determine who would hold mastery over the northern railway terminal at Juárez. According to British soldier-of-fortune, I. Thord-Gray, commanding a Villa artillery battery, the Villistas had 5,000 men while the federal forces numbered 7,000.

Thord-Gray described the Villistas as "brave little Indians and peons." They marched or rode, he said, without uniforms in shabby clothes. Some of them didn't have a gun but "proudly carried their *machetes* and vicious-looking long knives." The British officer remarked, "Scalping was not entirely out of fashion." Some of the soldiers "looked about 10 years old," he said. Villa ordered the *Soldaderas*, the famed women camp followers, to stay behind, but Thord-Gray later wrote, "Hundreds of them, hanging on to the stirrups of their mounted men," trudged along the dusty road.[11]

Some did more than wash and cook and comfort their men. Many picked up the guns of wounded soldiers and took their place in the firing line, fighting as bravely as any man. The *corridos*, the folksongs of the peasants, captured the spirit of these tough, courageous women, and one of the most popular, "Juana Gallo" tells of a young lady who was:

> *Always at the front of the troop you saw her*
> *Fighting like all the other soldiers*
> *In battle no federal soldier escaped her*
> *Without mercy she shot them with her big pistol.*

Early on November 23 the federal army attacked on a 15-mile-wide front in what became one of the largest and most fierce battles of the Mexican revolution. Federal cavalry attacks backed by intense machine-gun- and artillery fire went on well into the night. Each was bloodily repulsed by the heroism of Villa's troops with assistance from Holmdahl's blazing machine gunners.

On the following day, Pascual Orozco led 4,000 of his "Colorados" in an attempt to circle behind the rebels' left flank. Villa shifted his reserves and drove them back. Throughout the day, hundreds of terrified Mexicans in Juárez fled across the international bridge to El Paso, as the booming of the artillery shook window panes in the border cities. Trains full of Villa wounded began to return from the front until the Juárez railway station resembled the scene in *Gone with the Wind* where broken bodies covered the streets.

At 5 o'clock on the morning of November 25, Villa's troops took the offensive against a federal army exhausted after two days of futile attacks. A drive from the rebels' left flank didn't net the elusive Orozco, but, to Holmdahl's delight, isolated 2,000 troops under General Salazar and pinned them against the Rio Grande.

Villa sent a courier to the front with a command to take Salazar alive. "I will take him to the main square of the city and will have the pleasure of shooting him myself," he pledged. The canny Salazar, however, would not be taken, and many of his men escaped by swimming with their horses across the river while those on foot either swam or built rafts to float to the U.S. side. There they were rounded up and interned by U.S. cavalry patrols.

Salazar and Orozco escaped to the east across 200 miles of wild desert, leading a footsore caravan of 3,000 troops accompanied by hundreds of federal sympathizers, including many women and children. After a five-day trek the column, which straggled for 35 miles across the bleak plain, reached Ojinaga. After scattering the small Villista garrison, they seized the town and obtained food and precious water. They had left a trail of dead from Juárez to Ojinaga.

During the battle at Juárez, many El Pasoans, unable to sleep because of the incessant firing, spent the days and nights on their rooftops watching the fighting raging across the river. The spectacle became even more exciting when stray bullets whizzed over their heads. At the rooftop ballroom of the Paso Del Norte Hotel, there was a carnival atmosphere. Sedate couples interrupted their foxtrots to peer over the rooftop colonnade to watch when a particularly vicious storm of shooting drowned out the music.

Somehow, during the fighting, Holmdahl was interviewed by reporters whose dispatches appeared in major U.S. newspapers. Most credited Holmdahl with winning the battle, reporting that he "led charge after charge until the enemy was repulsed." The *San Francisco Call* added a touch of poignancy to the story, stating,

> *It fell to an American to display the most daring ability to fight under the fire of the enemy. Emil L. Holmdahl, now chief of Villa's artillery, is given the credit of the rebel victory and the holding of the Federals in check. Holmdahl is an Oakland man. While he is fighting as a soldier of*

fortune his white haired mother sits in her home at 617 Angar Street, Oakland, and anxiously awaits news from the Mexican border."[12]

Another newspaper quoted Holmdahl's mother, "I fear I shall lose my boy some day . . . he is so impetuous and so devoted to the cause of Madero that he will not be content to remain in the rear ranks."[13] The San Francisco paper commented, "This American, who is a major in the rebel forces, is the recognized strategist of the defenders of Juárez. In this morning's battle he displayed great fighting ability, and time after time led the charge against the federal positions."[14]

Although inferior in numbers, artillery, machine guns, and ammunition, Villa's wild cavalry attacks, covered by his machine guns and artillery, routed the federal troops. Many frightened federal soldiers, some only raw recruits, were found huddling together under a white flag. Villa ordered them shot to a man. In all, the federals had more than 1,000 killed, while Villa's forces suffered 200 dead and 300 wounded. During the battle, Holmdahl's Maxim guns did yeoman service in shooting fleeing government troops.

The aftermath of the battle was a bonanza for Villa, as his forces captured four trains loaded with supplies, several artillery batteries, a dozen machine guns, hundreds of rifles, and 400,000 rounds of ammunition. The battle of Tierra Blanca, however, "showed both the strengths and weaknesses of Villa's strategic thinking." On the positive side, he maintained the morale of his troops, his cavalry charges routed the enemy, and his position was well thought out. He had, however, "no concept of reserves . . . was not coordinating the battle . . . This was in the tradition of guerrilla fighting, but was contrary . . . to the command of a large army on the battlefield."[15]

A military expert later stated that Villa had no concept of how to use or defend against artillery. These weaknesses would manifest themselves in later battles, and, ultimately, would lead to the defeat of his proud army.[16]

A few days after the battle, Holmdahl led a patrol of forty mounted men through the desert southeast of Juárez, searching for a band of Huerta troops who were raiding Villa's supply lines. Based in the Texas border town of Ysleta, fifteen miles east of El

Paso, the band crossed the Rio Grande into Mexico on daring raids. Then they fled back to safety in Texas.

Holmdahl was tipped off to the whereabouts of the raiders by a U.S. army officer patrolling the north bank of the river. The officer may have been an old crony from the 20th U.S. Infantry Regiment now stationed in El Paso. They were patrolling the area in order to protect that border city from raids by the *bandidos* accompaning both armies. With this information, Holmdahl was able to slip up on the group's camp at dawn, and, although there were at least 200 federals, Holmdahl struck hard and fast.

By positioning his men between the Federal camp and the river, Holmdahl's surprise attack cut off the enemy escape route and scattered most of the band toward the Mexican river town of Zaragoza. Riding into town, Holmdahl was hit with a rifle bullet entering at the top of his shoulder near the base of his neck and coming out beneath the shoulder blade.

Knocked out of the saddle, Holmdahl fell into the dusty main street of Zaragoza. Laying there, he watched his infuriated men shoot many of the raiders out of their saddles and overrun and capture twenty-eight of them. Propped up against an adobe building, he saw the prisoners lined up against the wall on the opposite side of the street and promptly shot. For most, mercy was as rare a commodity as loyalty.

Holmdahl was taken to El Paso, where under the care of American doctors, he again demonstrated his recuperative ability. He granted an interview to a newspaper reporter, and he was pictured in a Chicago newspaper of December 13, 1913, looking fully recovered.[17] He was now thirty years old and had been wounded several times. There were streaks of gray in his hair and new lines around the eyes, but his thirst for action and adventure remained undimmed.[18]

In December 1913, Villa and his army rested and reequipped in Juárez. Villa realized he needed more guns and ammunition if he were to drive to Mexico City. To this end, he organized a massive smuggling operation in which Holmdahl became a key figure.

Setting up in the Sheldon Hotel in El Paso, Holmdahl made contacts with U.S. businessmen who, in exchange for cattle, cotton, copper, and silver appropriated by Villa, delivered guns and ammunition to the border. Since the beginning of the revolution, a great market in the purchase of "agricultural equipment" had suddenly grown in El Paso.

Hundreds of crates labeled as plows, harvesters, and windmills were delivered to El Paso by rail. From the freight station, they were unloaded at night into wagons hauled by mules and taken into the desert. There they were met by bands of smugglers who opened the boxes and transferred the cargo of Winchester .30-.30 carbines, Colt .45 caliber revolvers and cases of ammunition to pack mules. At night, in small caravans, they dodged the few American patrols and waded the shallow Rio Grande into Mexico.

Smuggling was big business in El Paso, and its leading citizens were often involved. The Sheldon Hotel was a rambling, four-story brick building in the center of downtown El Paso, and it was reputed to be the finest hotel on the border. It featured a gourmet cook and the best-stocked bar in Texas. In its paneled lobby Richard Harding Davis, John Reed, and other famous correspondents often swapped tall tales and more serious information with military men, such as Major Generals Hugh Scott and Frederick Funston. They were sometimes joined by the stiff-backed new brigadier John "Black Jack" Pershing.

After the Díaz forces were defeated in 1911, Madero, Orozco, and Villa held a victory dinner in the main ballroom. Later, after Villa fell out with Garibaldi, the former bandit swaggered into the hotel lounge, six-gun stuck in his belt, and announced "I'm going to kill that Italian bastard." It required all the peace-making abilities of El Paso Mayor Charles Kelly to persuade Villa that the hotel was not the proper place for an assassination. Soon after Garibaldi left Mexico forever.

A few years later the tables were turned during the Orozco revolt. Tracy Richardson smuggled a large pack train of guns and ammunition across the border to General Salazar who was then an Orozco ally fighting Villa. After the Salazar's men had used the arms to shoot up a Villa detachment, Villa posted a $10,000 in gold reward for anyone who would deliver Richardson to him dead or alive. After Richardson had disposed of several of Villa's failed bounty hunters, he learned the general and his side-kick Fierro were holding forth in the Sheldon Hotel bar. Backed by a few friends, Richardson entered the bar, pistol in hand. As the room suddenly became very quiet Richardson told Villa he had a choice; he could call off the reward or he could have his head shot off. Villa's face lit up in a broad beauteous smile. "Amigo," he said, "I call it off. Let's have a drink." Everybody, even Fierro, smiled broadly and the event was closed.[19]

At that time, Holmdahl was working as a Villa purchasing agent. He and Richardson undoubtedly sat in the spacious Sheldon bar, downed a few drinks, and swapped stories about the fine art of gun-running. And they both might have laughed when Sam Dreben changed sides and went to work for Villa. They probably didn't have to pay for many drinks themselves, as there were numerous arms salesmen looking for contracts, and a bevy of U.S. newspaper correspondents looking for colorful stories, who would gladly stand a few rounds of bourbon and water. If there were few contracts, there were stories aplenty.

The distinguished soldier and military historian, General S.L.A. Marshall, commenting on the times, later wrote, "Gunrunning was common along the border. A gunrunner was regarded as an adventurer, not a criminal." Marshall said that Holmdahl was "Villa's agent in negotiations with the business community in El Paso."[20]

Somehow, a deal must have gone sour, or maybe Holmdahl got a better offer. Or possibly, Holmdahl suffered pangs of conscience as he watched Villa and Fierro slaughter helpless prisoners. If Holmdahl had any remaining scruples, the last straw was an incident described by Marshall. Marshall recounted to the Institute of Oral History at the University of Texas at El Paso, that one evening he was in a Juárez bar and gambling joint called "El Gato Negro" ("The Black Cat").

There, Villa, in an expansive mood, pointed to a high Spanish comb worn by one of the dance hall girls. He bet Rudolfo Fierro that he could shoot the comb off of the girl's head without hitting her. Fierro bet him $25.00 he couldn't. Villa pulled out his big Colt .45, aimed, and fired. The girl, shot through the head, fell dead.

Villa with a sheepish smile, counted out $25.00 and handed it over to Fierro. It was considered quite a joke.[21]

On December 24, 1913, Holmdahl again wrote the adjutant general in Washington D.C. stating,

> Have just resigned as 1st Capt. of Artillery, with Gen. Pancho Villa's Rebel Forces in Chihuahua, my reasons for doing such; were on account of ill feelings and petty jealousies shown me by my superior officers."[22]

Maybe. Perhaps some of Villa's officers read the American newspapers and learned how the gringo had "won" their greatest victory. Maybe Holmdahl had mentally resigned from Villa's service but had not yet bothered to tell Villa.

The Washington letter also stated, "Can speak the Spanish language fluently. While campaigning through 13 [Mexican] states, I have learned the water holes, and trails." He gave as a reference Brigadier General Hugh Scott, the commanding general of U.S. forces along the border. In the letter, he added, "Before leaving Villa's forces, have taken a full list of all artillery and small arms."[23]

In supplying information to Washington, Holmdahl, now working for Carranza, could be said to be spying for three different armies. It was something of a record, even for the Mexican revolution. The gunrunning business was characterized by double-crosses, with more than a little *mordida* (bribery), highjacked shipments, unpaid bills, and failures to deliver paid-for arms. It was the latter that was probably partially responsible for Villa's raid on Columbus, New Mexico in March 1916.

In one case of chicanery during 1913, a "red-haired" con artist, whose sole capital was an extra collar and a clean shirt, ensconced himself in the Sheldon Hotel and convinced Villa supporters of his credentials as an arms dealer. With introductions from them, he met with Villa and announced he could "supply him with all the arms and ammunition he would need . . . at incredible low prices."[24] Villa then turned over $10,000 in American currency to the man.

After weeks elapsed without delivery, Villa crossed into El Paso and, after a brief search, located the man who was frolicking about the city's pleasure dens. At gunpoint he forced the man to refund the money. The redhead was very lucky that Villa cornered him north of the Rio Grande.[25]

By January 1914, Pancho Villa, in undisputed control of all Chihuahua, was the darling of the American war correspondents and viewed favorably by the American government. But underneath the surface—resentment was ready to boil over. Holmdahl was either lying to Washington in his December letter, or perhaps he meant he had resigned from Villa's forces and joined Carranza without telling Villa. Perhaps the *New York Times* got it wrong, for on June 20, 1914, the newspaper ran a story stating:

U.S. ARMY VETERAN TO LEAD REBELS INTO LOWER CALIFORNIA

Douglas, Ariz. June 19 — After the departure today of Major H.L. Holmdahl of Gen. Villa's personal staff from Agua Prieta for Nogales and Hermosillo, the statement was made by Constitutionalists, that he had been delegated by Villa to equip and lead an expedition to take Lower California for the insurgents. Such an attempt would require a march across the desert in order to capture Mexicali and Tia Juana. Three previous expeditions have failed.[26]

Aside from this brief dispatch, there is no other record of Holmdahl's expedition, if indeed it ever took place.

The disaster at Tierra Blanca and a series of other humiliating defeats broke the back of the Huerta regime, and in August 1914, the half-drunken general boarded a German ship and fled the country. The resulting vacuum of power left three major players on the revolutionary scene: Venustiano Carranza, the titular head of the revolution; Villa, whose Division of the North was the most feared fighting force in the country; and Zapata, who had triumphed in Morelos and parts of southern Mexico.

There had been growing conflict between Villa, the peasant, and Carranza, the *rico*. Aside from the economic and social gap, perhaps there was a generation gap between them. Villa was a vibrant, blunt, man of action in his thirties and Carranza, in his fifties, was cold, withdrawn, and intellectual. They were like people from two different worlds.[27]

> From the beginning . . . Carranza attempted to subordinate Villa to leaders he considered far more reliable and to limit his authority in Chihuahua and other territorieshe controlled.[28]

Villa at first tried to conciliate their differences.[29] But after repeated insults from the Carranza forces, he made a final break which led to "The greatest bloodshed in the history of the Mexican Revolution, and was also its most senseless episode."[30]

Zapata, interested only in his home province of Morelos, later withdrew from the competition, leaving Villa and Carranza to contest for Mexico. The two, more in a battle of egos than idealistic differences, cut relations and prepared for a showdown.

By December 1914, Villa was knocking on the gates of Mexico City, and on December 4, still allied with Zapata, the two met at Xochimilco on the outskirts of the capitol. Days later, he and Zapata moved their armies into Mexico City. Holmdahl, during this year of triumph for Villa, was secretly taking orders from General Benjamin Hill, Carranza's chief officer stationed along the Texas border.

Before they were at each other's throats, Gen. Alvaro Obregon, left, Pancho Villa, center, and Brig. Gen. John J. Pershing posed for this photograph. — Aultman Collection

☙ 10 ❧
Trial and Redemption

"I fight because it is not so hard as to work."
— rebel soldier to John Reed in
Insurgent Mexico [1]

As the Villa-Carranza split grew into open warfare, Holmdahl was commissioned by General Hill to spy out the location and strength of Villa's forces remaining in the north. How much military information was exchanged in the Sheldon Hotel Bar between the mercenary "friendly enemies" is not known. One suspects there was considerable cooperation and exchanges of information between the U.S. soldiers-of-fortune, whichever side they were on. Information leaks were constant, a fact that Holmdahl was to learn to his sorrow.

In October 1914, Holmdahl was ordered to organize a small army to operate behind Villa's lines in Chihuahua. He formed an alliance with Jorge U. Orozco, a Carranza diplomat who was formerly the Mexican Consul in El Paso. Also involved were José Orozco, a former colonel in the "Colorados" and a cousin of General Pascual Orozco, now in hiding somewhere in the United States, and Victor L. Ochoa, a Carranza agent.

Ochoa slipped back and forth across the border engaged in plots and counterplots. Unfortunately for Holmdahl, Ochoa was not very good at plotting. During the 1890s, he served three years in an

American prison after he was convicted of organizing a revolution against Díaz while on U.S. soil.

In 1911, involved in another plot against the dictator, he was caught, tried, and, convicted in a federal court. After serving eighteen months in jail, Ochoa was released at Carranza's request. After the revolution ended, on September 20, 1921, he was indicted for selling narcotics to an agent. To Holmdahl's later sorrow, he was up to his neck in the current plan.[2]

The Holmdahl *junta* contacted former Mexican army officers living in the United States who had previously fought with Díaz, Madero, Orozco, or Huerta. It really didn't matter. They and other volunteers, along with a boxcar loaded with military supplies, were to go by railway from El Paso, sixty-five miles to the west, and unload at the small cattle town of Columbus, New Mexico.

There they would dig up a secret arms cache in the desert that had been buried the previous year by the "Colorados." After picking up more local recruits, they planned to cross the border and rendezvous with Carranza troops in the area. The combined force would then capture the Villa garrison at Palomas, just across the border from Columbus. This action would cut Villa off from the west, while a force under General Hill attacked Juárez from the east.

Holmdahl, attempting to recruit a man named Frank Heath, stated, "I am organizing an army of 20,000 men to invade Mexico and take Juárez."[3] According to Heath's later testimony, Holmdahl said he held a commission as colonel in Carranza's army. If the invasion succeeded, Holmdahl said, it would be the death blow to Villa. Pancho relied on the Juárez-El Paso connection to sell the cattle, silver, cotton, and copper his "government" had appropriated to buy arms and ammunition for use against the growing strength of Carranza. Unfortunately for the *junta*, Heath was an undercover agent for the U.S. Immigration Department.

On October 15, Holmdahl received a telegram from an arms dealer in Galveston, Texas, stating:

> *We have option we believe on only stock thirty soft point Winchester cartridges in Texas option expires tomorrow do you care make us an offer*

on the entire lot of seventy five thousand we understand will be no further shipments this cartridge until after first year.
                                                     O.R. Seagraves MGR[4]

Whether or not Holmdahl purchased the ammunition is unknown, but he probably did because the .30-.30-caliber carbine was a popular weapon used by many armies during the revolution. Soft point bullets, however, had been outlawed under international law following the Geneva Convention of 1906. Soft points were forbidden because unlike steel-jacketed bullets, they expanded on impact and broke into pieces or tumbled through a soldier's body, creating a horrendous wound. During the Mexican revolution, however, when enemy wounded and prisoners were routinely shot without trial, the use of soft-point bullets was not a matter of concern to many people.

While there was much top-secret planning and many oaths sworn to maintain security, the plot, like most of its kind, was porous. One of the recruits later testified he was afraid of being killed by either side, so he spilled the plot to Héctor Ramos, head of the Villista secret service in El Paso. Ramos tipped off U.S. officials who stood ready to pounce.

On the night of October 31, 1914, several dozen hard-faced men were lounging about El Paso's Union Station. Victor Ochoa casually strolled among them passing out tickets for the El Paso and Southwestern train en route to Columbus, New Mexico, and Douglas, Arizona. Unknown to them, other eyes were watching. As the conductor bawled, "all aboard," the silent men filed onto the train, but the train did not start. Instead, burly men with guns drawn and badges pinned to their coats, shouldered their way through the passenger cars, arresting the volunteers. The men were agents of the U. S. Bureau of Investigation and customs agents.

The volunteers were herded into the railroad office and questioned. Most admitted they had signed up to fight for Carranza, more for his money than for his cause. Except for Ochoa, they were all released, for the American officers were after bigger game than a few penniless *vaqueros* hoping to join any army that would pay and feed them.

Meanwhile, Holmdahl was riding on a train carrying both passengers and freight, including a boxcar filled with military equipment labeled as agricultural supplies. For some reason, he received word not to unload at Columbus, but to proceed on to Douglas, Arizona, where he was to unload the merchandise, rendezvous with his troops, and cross the border near Agua Prieta.

As the train pulled into the Douglas depot, Bureau of Investigation officers arrested Holmdahl, routed his boxcar to a siding, and opened the crates. Inside they found 100 saddles, bridles and horse blankets, 75 cases of .30-.30 rounds, 50 cases of 7-mm carbine ammunition, 400 canteens, 160 .30-.40 caliber rifles, and nineteen boxes of other rifles. A box of bugles was also found.

Holmdahl, Ochoa, and several other plotters were taken before the federal district court in El Paso and charged with violations of the 1911 Federal Neutrality Laws, which forbade raising troops for foreign armies on U.S. soil. They were also charged with attempting to smuggle arms and ammunition across the border. Their penalty, if convicted, could be three years in a federal penitentiary and a fine of $10,000. After arraignment the men were released on bond pending a trial date.[5]

While out on bail, Holmdahl, with his usual boldness, continued his gun-running operations, as evidenced in a series of telegrams received from an arms dealer on December 12, 1914:

> *Major E.L. Holmdahl*
> *Can offer you salvage millimeters at thirty five per thousand under terms suggested by Brennan we have just turned down a cash offer of this amount giving Constitutionist (i.e. Carranza forces) preference can you use heavy pieces Gatling guns thirty forty Kraig [sic.] cartridges etc wire at our expense if you want us to write fully at Naco [a railroad depot on the Arizona-Sonora border] shortly will have best stock of war munitions in the south and it would be of mutual interest to keep in touch with us you ought to be able to use some of our army aeroplanes with experienced airmen furnished by us.*
>
> *Pierce Forwarding Co. 10:50 am* [6]

Presumably responding to an answer by Holmdahl, the company replied by Telegraph:

Major E.L. Holmdahl
Will only sell the millimeters subject to condition as comes from boat cannot guarantee salvage goods the market is good better wire acceptance immediately and arrange with your people for financial details as we can sell five times over at these figures.
Pierce Forwarding Co. Galveston. 1:58 pm [7]

Apparently the deal was settled as M. Brennan, a Holmdahl agent, telegraphed:

E.L. Holmdahl
As a favor got Pierce to let us have millimeters at same price as other offer they have opportunity to receive cash today if possible accept without guarantee and have (General Benjamin) Hill wire immediately guarantee of draft or COD.
M. Brennan Galveston 2:23 pm [8]

On January 10, 1915 Brennan telegraphed Holmdahl offering another deal:

Pearce Forwarding Co. Have fifteen hundred thirty rifles and carbines thirteen one hundred thousand forty five seventy Springfield cartridges forty fifteen hundred forty five seventy Springfield rifles ten.
M. Brennan [9]

Since this rather blatant negotiating was done over open telegraph lines, the parties either knew government agents were not monitoring telegraphic traffic or they were extremely careless.

In February 1915, Holmdahl was in Vera Cruz, probably illegally, since he was at this time out on bail and not allowed to leave the country. During 1915, U.S. mercenaries and Mexican revolutionaries regularly changed sides. It often resembled the western dance known as the "Paul Jones," where women formed a inner circle and men formed an outer. While the music played, men moved counter clockwise, women clockwise. When the music stopped, your partner for the next dance was the person in front of you. It didn't matter who it was so long as you continued to dance.

A prime example was a letter written by a Carranza brigadier general named Hernandez who on February 23, 1915, wrote El Paso Mayor Tom Lea:

> *Dear Friend and Brother.*
> *The bearer Major E.L. Holmdahl, is leaving (Vera Cruz) for your city to await trial by the U.S. Federal Court, accused of violating neutrality laws, the charges against him were made by Héctor Ramos, chief of Villa Secret Service, who has personal ill feeling towards the Major who was at one time connected with Villa as Chief of Artillery, leaving them to join our cause.*
>
> *The Major is a personal friend of mine, and I would greatly appreciate anything that you may do for him in receiving justice in pending trial. Wishing you every success in your new undertaking.*
> *Very Respectfully,*
> *J.H. Hernandez*
> *Brigadier General* [10]

Interestingly, if Mayor Lea was an ally in February, he had changed sides by December and was backing the *junta* of Pascual Orozco, Victoriano Huerta, and Inez Salazar, all formerly enemies. Now allies, they were planning an invasion from across the U.S. border.

According to statements made from a Federal jail in El Paso by six former Huerta officers, Tom Lea was in on the plot against Carranza. The officers stated they were part of a group of 200 recruits that had rendezvoused at Lea's El Paso ranch, where they were to be issued guns and ammunition and then would be joined by Inez Salazar. Salazar had been incarcerated in a New Mexico prison, but he broke out of jail and was riding to El Paso with fifty mounted and armed men who would lead the revolt.

The rendezvous at the Lea ranch was broken up when a troop of U.S. cavalry descended on the plotters. A score of volunteers were arrested, while the rest scattered and ran either into the desert and the surrounding Franklin Mountains or dived into the Rio

Grande and swam to Mexico. It is not improbable that Holmdahl, who had an informant in the "Red Flaggers" camp, was the man who tipped off the cavalry as to the time and location of the meeting. The six officers told U.S. officials that they made their statements because the *junta* failed to provide their families with funds, did not get them lawyers, and left them to rot in jail.[11]

The year of 1915 was also "The year of definition of the civil war with the defeat of the Villista and Zapatista armies."[12] It was also a year of hunger for many Mexicans. Amparo F. de Valencia who was ten years old that year recalled the suffering of her family:

> *"During the revolution I saw hunger. Sometimes one ate and sometimes one didn't.... There was no wheat because there was no harvest... Thus there was tremendous hunger all over... A big bakery had a big stack of sacks full of old hard bread that was sent to Sinaloa for feeding pigs... I bought a sack and dragged it home. The bread was full of ants and other things, but we felt happy.*[13]

Disease, she said, was another problem, "The victim would die, because there was no medical assistance."[14]

During most of 1915, Holmdahl's activities were shrouded in mystery. While awaiting trial, he continued working for Carranza as a spy, arms agent, and smuggler. He was not to surface again until October 14, 1915, when he and other plotters went to trial in El Paso's federal district court. It was a brief affair with little grounds for defense. Former Mexican revolutionary officers testified they were recruited and paid to cross the border and invade Mexican soil. Various arms salesmen testified that Holmdahl had bought and paid for weapons. A variety of American agents testified they had been approached by Holmdahl, Ochoa, or one of the Orozcos to join the "filibusters."

After a short time, the jury brought in a verdict of guilty against Holmdahl, Ochoa, and José Orozco. Jorge Orozco was found not guilty. It was the first case the government successfully prosecuted recruiters and gun-runners under the Neutrality laws. Because of that, or perhaps because rumor had it that many of the prominent businessmen in El Paso were involved in bankrolling the plot, Judge

Thomas S. Maxely showed leniency. The three were sentenced to eighteen months in a federal penitentiary and no fine was levied. After sentencing, the three were released on $7,500 bonds pending appeal.[15]

While out on bail, Holmdahl learned of the treachery of Tomás Urbina, an old compadre of Villa's during his bandit days. After being badly beaten by Carranza forces, "General" Urbina had become a deserter. Abandoning his shattered forces, the old bandit took his accumulated loot, said to be worth millions in gold and silver, and fled to his stronghold, Las Nieves, in Durango.

Smelling betrayal, Villa took Fierro and 200 men, and rode to Urbina's fortified hacienda. Rushing the gate, they shot their way into the compound and wounded the bandit general in the leg. At first, Villa seemed moved at the sight of his old friend and chatted amicably with him, while Fierro salivated with eagerness to shoot him.

Fierro's men, meanwhile, searched the *hacienda* grounds and questioned, not too gently, Urbina's surviving men. They quickly located gold and silver bars that had been dumped down a well. Hauling it up, they dumped it at Urbina's feet. Villa's benevolence abruptly ended. "Shoot him," he commanded. Then he mounted his horse and led his men from the *hacienda* at a trot.

Fierro stayed behind. Stories had it that he shot Urbina in his good leg and in both arms before shooting him in the belly, and smoked a long cigar while watching Urbina's face turn ashen. Loading the gold and silver in his saddlebags, he mounted and spurred his horse to catch up with Villa.

Fierro had been too eager, for there were rumors of other caches of treasure buried by the crafty old bandit. When Holmdahl heard them, he filed them away in his memory. Later, when things calmed down, perhaps he would make a little trip to Las Nieves.

If nothing else good came from Urbina's betrayal and subsequent murder, it was that Fierro, his saddlebags still loaded with bullion, attempted to cross a flooded field when his men balked. The field was full of quicksand, they feared, and it was too dangerous to cross on horseback. Calling them cowards, Fierro boldly rode out into the field and immediately plunged into quicksand. Then his horse stumbled and fell. The horse and rider began to sink, both pulled down by the heavy bags of gold and silver. Cursing, Fierro shouted, "Throw me a rope."

His men stood silent and unmoving at the edge of the field, watching Villa's butcher sink, until the quicksand sucked him all the way under. Soon, the only trace of Rudolfo Fierro was his broad-brimmed sombrero on the surface.

Meanwhile, in what might be termed another masterpiece of *sang-froid*, a little more than a month after his conviction, Holmdahl applied for a commission as an officer in the United States Cavalry. On December 29, 1915, he filled out a three-page government form addressed to the adjutant general of the U.S. Army. On the application he stated he held the rank of colonel of cavalry with the Carranza forces and was formerly chief of Artillery under Villa.

To endorse his application, he gave a list of references, including Tom Lea, mayor of El Paso; Lee Hall, chief of police of El Paso and a former Texas Ranger; a banker from Morenci, Arizona; a captain in the U.S. Army stationed at Fort Bliss; and to top it off, General Hugh Scott, chief of staff and commanding officer of the U.S. Army. He gave his address as a post office box in El Paso.[16]

While the captain and the general may have appreciated Holmdahl's military experience, one might wonder if there was a connection between the mayor, the chief of police, and the banker. Whether the banker supplied money for the filibusters, while the mayor and the chief of police looked the other way may never be known.

On March 28, 1916, the war department answered Holmdahl's application by stating that he failed to qualify for appointment as an officer of volunteers because of regulations stating that "no applicant is eligible for appointment as second lieutenant who is more than 30 years of age." Holmdahl was thirty-two years old. If the war department knew that their applicant was a convicted felon, it was not stated.[17]

Fate, however, intervened, when Holmdahl's old boss, Pancho Villa, galloped into Columbus, New Mexico, on March 9 with a band of 400 men, shot up an army encampment, burned the town, and killed sixteen Americans. The rage and frustration that led Villa to attack an American town lies in the complicated relationships between the two countries and the tangled loyalties of both Mexicans and Americans.

In 1915, after five years of warfare, the economy of Mexico was shattered by marauding armies which stripped and devoured anything they could steal, sell, or eat. U.S. financial and business interests had invested billions of dollars in the country under, what was to them, the benevolent rule of Porfirio Díaz. Americans owned some of the richest farm and grazing lands, had controlling interests in and operated many of the railroads, had majority interests in many of the mines, and with British companies, dominated the Mexican petroleum industry.

As the devastating revolutionary warfare continued, American financial and commercial interests put increasing pressure on Congress and the president of the United States to take action to stop the fighting and bring peace and stability back to Mexico. One Congressional group, led by Senator Albert Fall of New Mexico, advocated sending the U.S. army into the country to end the fighting. Fortunately, both President William Howard Taft and his successor, President Woodrow Wilson, rejected the idea.

Wilson, however, decided the United States government could aid in stabilizing the country by extending official recognition to one of the two contending parties as the legitimate government of Mexico. The United States had refused such recognition since Madero's murder in 1913, and the dilemma facing the president was who to recognize, Villa or his deadly enemy Carranza?

For years Villa had been considered a friend of the United States. He had been scrupulous in protecting American lives and property and when, in 1914, U.S. forces seized the Mexican ports of Tampico and Vera Cruz, he gave his tacit consent. He stated that the U.S. Navy was justified in its actions after American sailors had been falsely imprisoned by Mexican troops. In addition, Villa had a long standing-personal friendship with General Hugh Scott, who became the Chief of Staff of the U.S. army. He felt confident that he would be the one favored by the U.S. government.[18]

Carranza, on the other hand, had always been vocally anti-American. When the two Mexican ports were seized, he threatened war with the United States and also secretly gave support to the Plan of San Diego. This movement planned to conquer Texas, New Mexico, and Arizona and set up a independent government allied with Mexico. Far-fetched as the plot seemed, it was first supported by Huerta, then received help from Carranza and financing by the German secret service. Members of the Plan carried on brutal raids across the Texas border in 1915. With World War I raging, the Germans hoped to embroil the U.S. in a war with Mexico so as to divert aid intended for the Allies.

The raiders killed a number of American ranchers and murdered a crew working on a border irrigation project. They ambushed a U.S. cavalry trooper, cut off his head, put it on the end of a pole, and with their gory trophy, paraded back and forth along the south bank of the Rio Grande. One night they derailed a train near Corpus Christi and stormed through the passenger cars shooting Americans. When the U.S. government complained to Carranza, who was nominally in control of the area, he replied that if only he was recognized as the legitimate head of the Mexican government, he would, with this increased prestige, be able to exert the authority to end the raids. There was the scent of blackmail in the Carranza response.[19]

The American business community, however, taking a second look at Villa, pointed out that the former bandit, semi-literate at best, could not possibly form a stable government that could bring peace and renewed prosperity to Mexico. His economic beliefs, if any, were uncertain, and he had a nasty habit of shooting people on a whim. Also, he had just lost three battles to the Carranza forces.

For all his anti-American rhetoric, Carranza, whose forces were now in control in Mexico City, understood Mexican bureaucracy. He had the support of many affluent and middle-class Mexicans and came from a family of wealthy landowners. American business interests believed they would be safe under a Carranza regime. On October 15, against the protests of General Hugh Scott, President Wilson recognized Venustiano Carranza as the head of the legitimate government of Mexico.[20]

Villa was not only enraged at the affront, but he was hurt financially. Betrayal by his American friends was bad enough, but he was now relegated to the status of a bandit, and all purchases across the American border were prohibited. Carranza controlled all the Mexican ports on both coasts. Land-locked in Chihuahua and Sonora, Villa had no ready source to supply his army.

In late October, Villa decided to cross the Sierra Madre Mountains from his stronghold in Chihuahua into Sonora and capture the town of Agua Prieta, just across the border from Douglas, Arizona. There, at the Mexican Custom House, he would take a large financial bite out of the traffic in cattle and copper ore imported into the United States at the railhead in Douglas. With the money, he planned to set up smuggling operations along the sparsely settled borderlands and thus supply his troops.

Carranza, getting wind of the attack, received permission from the U.S. Government to reinforce his garrison at Agua Prieta by using the American railroad that ran from El Paso to Douglas. From Juárez, he sent troops, machine guns, artillery, barbed wire, and three very large searchlights into El Paso. There they were loaded onto an American train, and, within a day, arrived in Agua Prieta.

Villa's army reached Agua Prieta on November 1, 1915. Surveying the town's defenses, Villa decided to launch his *golpe terrífico* at midnight so that in the confusion of darkness he would overrun the federal forces. As his troops galloped toward the enemy trenches, however, the three big searchlights, with current supplied by the Douglas Power and Light Company, lit up the battlefield. Villa's cavalry became easy targets for the federal machine gunners.[21]

The charge was shot to pieces, and Villista troops were more than decimated. At Agua Prieta, a military expert proclaimed, "Villistas learned that an assault against a position covered with barbed wire, defended by cross firing machineguns, supported by artillery firing high explosive, is doomed to failure."[22] Slinking back across the mountains with the survivors of his once proud army, Villa swore revenge against the Americans who he believed betrayed him and who were the cause of all his problems.

At the end of 1915, Villa had "only a few hundred followers. . . left of an army that had once encompassed between 30,000 and 50,000 men . . . His popularity among the civilian population in Chihuahua had reached an all-time low."[23] On March 9, 1916, Villa had only 450 men and was low on ammunition and food. His horses were worn down and his cause was flagging. At four o'clock that morning, his men rode into the New Mexican border town of Columbus burning and killing.

Historians have speculated for years about Villa's motives. Some say it was a desire for loot, and that he hoped to steal guns, ammunition, and fresh horses from the American garrison there. Another reason advanced was that a Columbus merchant, Sam Ravel, had taken money from Villa to purchase arms. Ravel swindled Villa, keeping the money but failing to deliver the weapons promised, and Villa hoped to capture Ravel and give him a lesson in business ethics.

Still another version had it that Villa hoped by raiding U.S. soil to force the U.S. army to invade Mexico. By fighting the gringos, Villa would again be a hero to all Mexicans. One theory has it that the raid was motivated by pure revenge and hatred of the Americans, who, he believed, had betrayed him by recognizing Carranza and causing him to lose the battle of Agua Prieta.

A wilder and more bizarre theory held that American businessmen paid Villa to attack Columbus, hopefully causing the United States to attack and occupy northern Mexico to their financial advantage. Then there was the version where the German secret service, supporters of the Plan of San Diego, wanted the United States to engage in a war against Mexico. Then, the United States would use munitions from its armaments industry in that war instead of shipping them to Great Britain and France.[24]

Columbus, with less than 400 inhabitants, boasted a railroad depot, two small hotels, a few stores, and several scores of scattered adobe houses. It was garrisoned by 300 troopers of the U.S. 13th Calvary Regiment encamped a few hundred yards from the town. When the Villistas attacked only a few sentinels and the kitchen staff, preparing the soldiers' morning meal, were awake.

Villa directed half of his men to attack the center of town and the others to hit the army camp, stealing weapons and horses. Screaming "Viva Villa" and "Muerte a los gringos" ("death to the Americans"), they stormed into the Commercial Hotel, dragged five half-asleep Americans from their beds, and shot them to death. Mrs. J. J. Moore saw her husband and her infant child torn from her arms and shot. She was raped, then shot and left for dead, but she survived.

In a vain hope of escaping the rampaging Villistas, Milton James half-carried his pregnant wife from the Hoover Hotel, but the raiders spotted them, opened fire, killing her. Other Americans ran into the desert, hiding in nearby cactus-filled *arroyos* and ditches, as the entire commercial block in the center of town went up in flames.

The raiders had less luck attacking the 13th Calvary. When a dozen of them burst into the cooks' shack, they were doused with boiling coffee. One was decapitated by an enraged baker swinging an axe, and the rest were assaulted by cursing cooks swinging meat cleavers and army-issue potato mashers, which could crack a skull like an eggshell. One raider was brained by a trooper on kitchen patrol swinging a baseball bat like a berserk Babe Ruth.[25]

Troopers grabbed rifles and quickly set up machine guns, blasting away at raiders outlined against the flames of the blazing business district. Columbus became too hot for the raiders as the sky began to lighten with the coming dawn. Grabbing anything they could carry, they scrambled to their horses and retreated across the border. As a bugler blew "Boots and Saddles," troopers, some still in their underwear, grabbed Springfields, bandoliers of ammunition, and their saddles, and rushed to the stables.

Major Frank Tompkins, bleeding from a Mauser bullet that raked his knee, led his men in a headlong pursuit of the fleeing Villistas. On fresh cavalry horses they outran many of the raiders and shot them down during a fifteen-mile chase. Then, low on ammunition and with no water, Tompkins, called a halt and the troopers rode back to Columbus.[26]

Returning, they counted seventy-five Villistas lying dead on the road while the bodies of more than sixty raiders lay dead in the streets of Columbus itself. The bodies in Columbus were heaped in a pile and burned, while Villista corpses on the road were left for the buzzards. Seventeen Americans were killed: nine civilians and seven troopers of the 13th Calvary.[27]

The United States was quick to respond. Brigadier General John J. Pershing was named to command a U.S. army ordered to drive into Chihuahua and smash Pancho Villa. Secretary of War Newton D.

Baker wired Pershing that he was authorized to employ whatever guides and interpreters were necessary. If the War Department didn't want Holmdahl's services, one "Black Jack" Pershing needed him badly.

Pershing may well have known or heard of Holmdahl, either during his Philippine campaigns or during his service along the border. An experienced soldier, Pershing graduated from West Point in 1886. He took part in the last campaign against the Apache Indians, and was cited for gallantry fighting alongside Teddy Roosevelt and his Rough Riders at San Juan Hill during the Spanish-American War.

Pershing taught tactics at West Point and was later assigned to the 10th Cavalry, composed of black soldiers, known as "Buffalo Soldiers." From this command he acquired his lifelong nickname of "Black Jack." In 1899, Pershing was ordered to the Philippines, where on Mindanao and Jolo he fought his first campaigns against the Moros. In 1901, as a captain, he launched four more major expeditions against the troublesome warriors. In late 1903, he married Helen Frances Warren, the daughter of Francis E. Warren, a powerful U.S. Senator from Wyoming.

After distinguished service as an observer on the Manchurian front during the Russo-Japanese War, he was promoted to brigadier general by his old San Juan Hill comrade, Theodore Roosevelt, now President of the United States. After further service in the Philippines, Pershing was ordered back to the United States. When the usual malcontents carped that Pershing's promotion over the heads of many senior officers was a result of political pull, President Roosevelt reported, "To promote a man because he married a Senator's daughter would be an infamy; to refuse him promotion for the same reason would be an equal infamy."[28]

On January 20, 1914, Pershing was transferred to Fort Bliss near El Paso, with the responsibility for supervising U.S. units patrolling the border. His wife, three daughters, and a son were left behind in quarters at the army base at the Presidio at San Francisco. Shortly before he planned to bring his family to new quarters at Fort Bliss, on August 20, 1915, he received news that haunted him the rest of his life. A fire had broken out in his family's quarters, and his wife and three daughters were burned to death. Only his small son was survived.

An austere man, he walked ramrod-straight, and if he was overcome with grief, it was not apparent from his square-jawed, stiff-lipped expression. Pershing was a sad, lonely man after this tragedy, and often he wandered into the desert, followed by several *mariachis*. He sat on the rocky ground while the men played "La Paloma" ("The Dove"), his wife's favorite song, and reflected on the needs of his forces.

Most of all, he needed men who knew Mexico, its terrain, its language, and its people. Holmdahl, who had been fighting over that terrain for six years, was quickly hired as one of the civilian scouts who knew the land, the language, and the tactics of the Villista guerrillas.

As Colonel H.A. Toulmin, who rode with Pershing, wrote,

> *The terrain is a desolate, barren, sandy plain with rolling foothills—blazing tropic heat to cold snows, (and) hostile winds and snowstorms. The Pershing Expedition placed its reliance on guides, cowmen of the ranges, half-breeds, ranch bosses, adventurers who had fought either against or for Villa, gunfighters, gamblers—the remnants of the old Indian frontier.*[29]

By the end of March, convicted felon or not, former Sergeant Emil Holmdahl was again riding with the U.S. Army.

Map 4: *The general campaign route of the punitive expedition.*
— Halcyon Press

11
Young "Blood and Guts"

"Poor Mexico, so far from God and so close to the United States."
— Porfirio Díaz

The punitive expedition, more than 10,000 strong, invaded Chihuahua on March 16, 1916. Pershing had three cavalry regiments, two infantry regiments, a battalion and two batteries of field artillery, plus the usual support units. His force included the army's First Aero Squadron, consisting of eight rickety airplanes. The expedition entered Chihuahua in two columns, one coming from the railhead at Columbus and the other further west from Culberson's Ranch in New Mexico. The two columns met about 100 miles south of the border at the American Mormon settlement of Colonia Dublán.[1] There, Pershing hired Mormon guides, whose knowledge of the countryside proved invaluable.

In addition, it was hoped that the army's aircraft could spot Villa's troops and report back swiftly on their tactical maneuvers or escape routes. It was, however, to be a vain hope.

If Pershing hoped that his aircraft could emulate those of the European powers which routinely carried out bombing and strafing missions, spotted for artillery, photographed enemy positions, and reconnoitered their troop movements, he was to be bitterly disappointed.

The ill-fated First Aero Squadron, Aviation Section, Signal Corps, consisted of eleven officers, eighty-four enlisted men, eight Curtis JN-2 "Jennies," ten trucks, and a touring car. The aircraft were not supplied with accurate maps, reliable compasses, machine guns, bombs, or reliable engines. The "Jennies" were underpowered, had inadequate control systems, and were usually unstable in any but straight, level flight.

A troop of the 7th Cavalry, "Custer's Own" departs Ft. Bliss near El Paso to join the punitive expedition. — Aultman Collection

Although the squadron's commander, Captain Benjamin Delahauf Foulois, and his brave and capable pilots repeatedly risked life and limb to carry out reconnaissance missions, the obsolete aircraft proved useless. Their sole capability proved to be as messengers carrying dispatches to widely separated units and occasionally in locating stray U.S. cavalry.

On the squadron's first reconnaissance mission, which was to aid cavalry units hunting Villistas, Foulois had to report failure. The plane's ninety-horsepower engine lacked the power to take it over the peaks of the Sierra Madre Mountains.

Further flights resulted in a variety of equipment failures, navigation errors, abandoned aircraft, and teeth-jarring crashes, so that within a month, six of the eight planes were reduced to junk scattered over the Chihuahuan desert, and by the end of April the squadron was no longer operational.

In June, however, four new 160-horsepower Curtis R-2s arrived at Pershing's headquarters. Sadly, assembled hurriedly, they were missing essential parts. Pilots found their laminated wooden propellers often came unglued in the dry heat of the Mexican desert country.

After suffering many crashes and painful injuries, surprisingly, none of the pilots were killed. But, by August 1916, the First Aero Squadron came to its inglorious end. All the more need for men like Holmdahl who knew the enemy country.[2]

The expedition's main problem was not Villista troops but supplies. Villa realized he was too outgunned, outnumbered, and under-supplied to meet tough American regulars in conventional battle. Also, his troops were not the tightly disciplined fighting men of the Division of the North, as many were untrained youths drafted from small villages raided by Villa. They rode with him only because they had the choice of joining his band or being shot on the spot. Avoiding fights whenever possible, his men scattered into small bands, hiding out in the vast wastelands of Chihuahua.

The Americans were prohibited by the Mexican government from using the Northwestern Railway which bisected western Chihuahua. To supply the expedition they would have to depend on the inadequate Mexican road system. To sustain itself in enemy country, the expedition daily needed 100 tons of supplies to feed and equip the troops and 110 tons of forage to keep horses and mules healthy. Water was always a problem in this desert region.

It soon became obvious that the horse- and mule-drawn wagons of the army could not possibly supply the expedition over the broken, rock strewn trails called roads in Chihuahua. General Hugh Scott, now chief of staff of the army, quickly purchased more than 600 trucks. He had to hire civilian drivers and mechanics to operate the motor fleet, as these trades were virtually unknown in the army. For staff use and reconnaissance, he purchased a dozen Dodge touring automobiles.

In the searing heat of the desert and chilling cold of the mountains, many breakdowns occurred, but the civilians managed to keep supply lines open. Supplies were always tight, but there was enough to enable the expedition to probe 250 miles deep into Mexico. As important as the transport were the civilian scouts who knew the water holes, grazing lands, and trails since hired Mexican guides proved unreliable. Often they led cavalry patrols into the mountains on wild goose chases. The population was hostile and merchants usually refused to sell supplies to the Americans, not that they loved Villa—many despised him—but they hated the presence of foreign soldiers on Mexican soil.

For the first two months, Holmdahl's activities are unknown. The reason for this void is that many records of intelligence operations conducted during the expedition are missing. It is possible they were destroyed in the 1920s, on the orders of General Pershing, then Chief of Staff of the United States Army.

Brig. Gen John "Black Jack" Pershing was frustrated by the limitations placed upon his punitive expedition and thwarted in his attempts to kill or capture Villa. — National Archives

There is evidence that Pershing hired two Japanese cooks to poison the elusive Villa but the mission failed. In those naive days assassination was not considered "cricket," so it is highly likely that when Pershing became head of the army, he purged the files of all relevant documents. This included any reports relating to other assassination plots and intelligence operations.[3]

This purging of documents gives additional credence to a statement made by Holmdahl in a 1962 interview with historian Bill McGaw. Holmdahl told McGaw that sometime in 1917, after Villa had successfully eluded the expedition's efforts to capture him, he was ordered to report to the office of General Bell, commanding officer at Fort Bliss in El Paso. According to Holmdahl, General Pershing and one of his officers, Colonel Herbert Slocum, com-

The U.S. cavalry, the backbone of Pershing's force, probed 250 miles into the desert and mountain country of Chihuahua. — Aultman Collection

manding the U.S. 13th Cavalry Regiment, gave him three $20.00 gold pieces for expenses to travel to El Paso.

At Fort Bliss, Colonel Slocum met with him in General Bell's office and offered him $100,000 to go back into Mexico and kill Villa. Slocum stated that payment would not be from government funds. The $100,000 would be paid by Russell Sage, Slocum's millionaire father-in-law. Holmdahl said he turned down the offer for three reasons. First, he said, "I liked Villa personally, even though I fought against him." Second, he said, "I am not a assassin." Third, he admitted, "If I killed Villa, I never would have gotten out of Mexico alive." Holmdahl's career suggests the third reason seems the most creditable.[4]

While many of the exploits of Pershing's civilian scouts are lost to history, a few have been recorded. Art Leibson, the biographer of Sam Dreben, recounts how that daring mercenary, disguised as an itinerant Syrian peddler, hitched up a wagon-load of miscellaneous goods and traveled through Villista country. During his travels, he picked up valuable information on guerrilla movements, which he sent back to Pershing's headquarters.

His adventures nearly came to an abrupt end in a small village, when he was recognized by a Villista with whom he had previously soldiered. When the man gave the alarm, Sam leaped on his horse and galloped away, outdistancing pursuers who would surely have shot him if they had caught him. During his escape, to the delight of the villagers, he abandoned his wagon load of pots, pans, bolts of cloth, and various colored ribbons. It was probably the only bonanza the poor villagers ever received from the revolution.[5]

Villa continued to elude the Americans; however, he was shot in the leg during a skirmish with troops loyal to Carranza. In great pain, the general was carried by his "Dorados" to a cave in the Sierra Madre mountains. Later, some of his men recalled, a troop of U.S. cavalrymen camped on a road directly below the cave entrance.

Sitting around their campfires, in the chill of the mountain evening, the Americans sang repeated choruses of the British Army's favorite marching song, "It's a Long, Long Way to Tipperary." Villa, listening, with little knowledge of English, thought the words sounded like the Spanish, *"Se jaló el buey con tapadera,"* which he translated as, "The ox got drunk with blinders on." "A strange song," he muttered to his *amigos*, "I wonder what it means"?[6]

In mid-April 1916, the expedition made its deepest probe, reaching Parral, 250 miles south of the U.S. border. Near the border of the state of Durango, the town was to be the scene of a number of events in Mexican history that were both bloody and bizarre. On April 12, after receiving permission from local authorities, Major Frank Tompkins, commanding two squadrons of the 13th Cavalry, rode to Parral with an advance party of eight officers and twelve enlisted men.[7]

While Tompkins and an aide talked with the *alcalde*, the rest of the party lounged in the city square. Suddenly a crowd formed, led by a beautiful, blonde, green-eyed German woman named Elsa Griensen. The woman, waving her arms and speaking in a strident voice, whipped the crowd into a frenzy of hatred against the Americans. She challenged their manhood and urged them to attack the lounging cavalrymen.[8]

The mob advanced on the Americans throwing horse dung and rocks, screaming "Viva Villa" and a varied array of Mexican curses. Tompkins rejoined his men and, rather than fire into a crowd of civilians, he ordered a withdrawal. The cursing mob, led by the green-eyed blonde, followed.

They were joined by Mexican soldiers loyal to Carranza rather than Villa. They, however, opened fire on the Americans, shooting a sergeant through the head and killing him instantly. After seven more cavalrymen were shot, one fatally, Tompkins had had enough and ordered his men to return fire. Crack shots with their deadly Springfields, the troopers of the 13th killed thirty-two persons, wounding twenty-five. Tompkins then ordered a withdrawal.[9]

In May 1916, Holmdahl again appeared on the scene, improbably involved with a young subaltern named Patton. After two months in Mexico, Second Lieutenant George S. Patton, Jr. was spoiling for a fight. Although an aide to Brigadier General John J. Pershing, he had been sent on many a dangerous ride, carrying orders to different columns of cavalry searching the Chihuahua desert for the fast moving Villistas. When the punitive expedition was formed, Patton badgered Pershing for a job as aide-de-camp. When he got the appointment, he became the most eager beaver on that demanding officer's staff. Thirty years old, tall, slender, aristocratic and good-looking, he was popular not only because of a likeable personality, but also because he would take on any dirty or difficult job with determination and enthusiasm.

By May, Pershing had established his headquarters near Rubio more than 200 miles deep in Chihuahua. Feeding men and horses in that barren countryside was more difficult than fighting the occasional skirmishes with stray Villistas. Since that part of Chihuahua could hardly feed its own population, it could much less provide fodder for the *gringo* invaders.

When intelligence sources reported that corn could be purchased at nearby Rubio and Coyote, Pershing called on his all-purpose aide, Second Lieutenant Patton, to take a small party of three automobiles and ten soldiers on a corn-buying expedition. Before departing camp on May 14 in three big Dodge touring cars, Patton surveyed his little band. There were two civilian drivers, ten infantrymen armed with the famed Springfield Model 1903 rifles, and two civilian scouts. One, a tall, lanky man, was Emil Holmdahl, who leaned against one of the cars with a saturnine grin and a big Colt .45 strapped to a gunbelt.

Holmdahl, by then, was a legend among the war correspondents and mercenary soldiers who had been reporting on or fighting in Mexico since 1910. Perhaps Holmdahl's grin was in anticipation that the young, and as yet combat-untried lieutenant would order his men to carry sabers. For if there was one flaw or eccentricity in Patton's makeup as a soldier, it was his dedicated belief in the value of charging with old-fashioned cavalry sabers. Patton had, in fact, designed the saber issued to U.S. cavalry regiments and constantly preached a doctrine of the *arme blanche* to anyone who would listen.

Most officers, including Pershing, only smiled at the young officer's declamations. They often pointed out that a line of cavalry each firing seven shots from the recently issued Model 1911 Colt .45

Second Lt. George Patton got his nickname, "Blood and Guts," after he brought three bloody corpses into Pershing's camp. — Aultman Collection

caliber automatic pistol had much more shock power than troopers leaning from their saddles trying to stick or slash an opponent. That morning, however, there was no mention of sabers, and the little group climbed into the three cars and drove twenty miles down the bumpy trail to Coyote. There Patton bought a few bushels of corn. Next they drove a few miles further to Rubio.

Entering the town a few minutes before noon, Holmdahl spotted a number of men loitering around the plaza. Although they were unarmed, he recognized some of them as Villistas he had soldiered with in campaigns against Huerta. "They are Villa's men," he whispered, "and they are a bad lot." As the men sighted Holmdahl, they drifted away down the crooked side streets of the town. Holmdahl's warning, however, set off alarm bells in the young lieutenant's mind. Colonel Julio Cárdenas, former leader of Villa's elite troop of "Dorados," was rumored to be in the area.

One of the Dodge touring cars used by Patton in his raid on the Rubio ranch. It was the young officer's first gunfight. — Aultman Collection

Patton remembered that twelve days earlier on May 2, he had accompanied "H" Troop of the 11th Cavalry on a swift approach to the San Miguelito Ranch, eight miles north of Rubio. There the wife and mother of Julio Cárdenas were living in a large *hacienda*. At that time, the troop deployed around the *hacienda* and swooped down upon the walled main house only to find it deserted. But Patton had a premonition about the place and he studied the buildings and terrain, just in case he might return there another day.

That morning, he reasoned, if Villa's men were in Rubio then Julio Cárdenas might be holed up at the ranch. Driving a few miles north of Rubio, Patton called a halt and briefed his band of fifteen men. The *hacienda* was built in two L-shaped wings with a walled courtyard encompassing the entire structure. There was a horse corral a few dozen yards beyond the main gate.

Windows in the wall faced out, but there was only one gate from which a horseman could ride out and escape. The *hacienda* was 200 yards east of the road that Patton had followed from Rubio, and on the side opposite the road was the main gate. If the Villistas attempted to escape, they would have to gallop out of the gate, cross the road, and ride for the mountains to the west.

Patton's plan was for the first auto, in which he rode, to drive past the *hacienda* and then make a quick stop. Patton and two men

(From a sketch drawn by Lt. Patton)

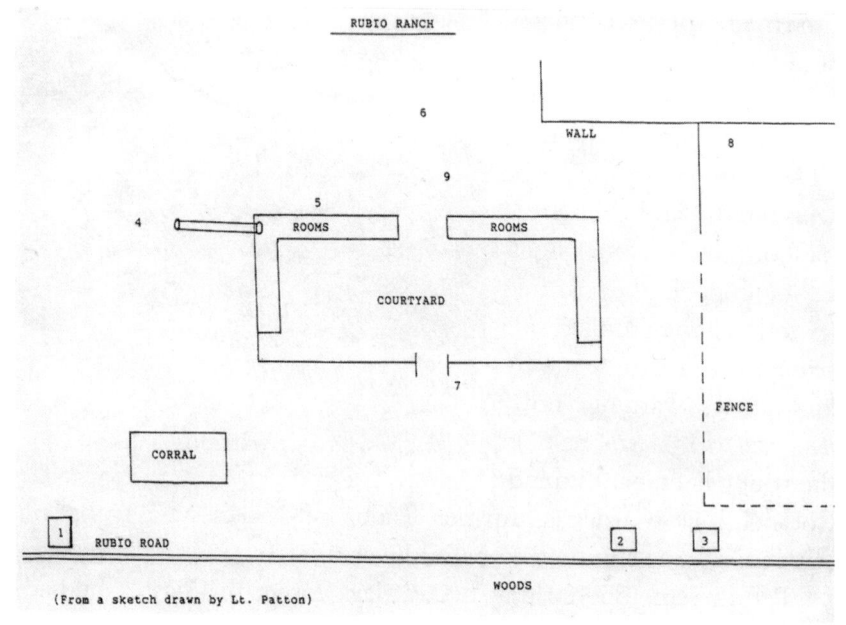

Map of the Rubio Ranch from a sketch by Patton.
1. Patton's auto
2. and 3. Other autos
4. Dead tree where Patton climbed the roof
5. Bandit on horseback killed
6. Bandit on horseback killed
7. Col. Cárdenas leaps from window
8. Col. Cárdenas killed
9. Cow skinners

— from a drawing by Lt. George Patton

would leap out of the auto and run to the northern end of the structure. Holmdahl and the driver would remain in the car and cover the north side of the walled area near the corral.

The second and third cars were to stop on the road south of the complex. Three men from each of those cars would dismount and dash to the southern end of the building. They would intercept anyone trying to leap from the windows. The two men remaining in each of the cars would be able to stop others running from the building and, if necessary and terrain permitting, to drive in pursuit of any fleeing bandits.

As the three cars drove up by the ranch buildings, Patton spotted three elderly men and a young boy skinning a cow in the front yard. The boy eyed the braking autos, turned, and ran through the front gateway waving his arms and yelling. Seconds later he returned and calmly went back to helping his elders skin the cow.

Patton said he "jumped out [of the auto], rifle in my left hand" and ran to the northern end of the hacienda, while the men from the other cars rushed to the southern end of the complex." He later wrote, "When I was about 15 yards from the gate three armed men dashed out on horseback . . . I drew my pistol and waited to see what would happen if they were Carranzistas."[10]

Patton yelled, "Halt." Hearing his yell, the riders wheeled their horses and galloped directly at Patton and his men. Patton later recounted that when the horsemen were about twenty yards away, "All three shot at me, one bullet threw gravel on me. I fired back with my new pistol, five times." Two of the bullets found their mark, one hitting the first rider in the arm and another rupturing the belly of the other horse.

At the sound of gunfire, Holmdahl, on the other side of the building came on the run, shooting rapidly. To get out of the line of friendly fire, Patton leaped around the corner of the complex, as three Villista bullets, possibly fired from one of the windows, smacked the wall by his head and showered him with adobe dust.

As the young lieutenant caught his breath and reloaded his pistol, he spied Holmdahl and one of the drivers sprinting toward him. Patton swung back around the corner of the wall and recalled, "I saw a man on a horse come right in front of me, I started to shoot at him but remembered that Dave Allison had always said to shoot at the horse of an escaping man and I did so."[11]

Dave Allison, a noted Texas law officer and range detective, had met Patton and exchanged a few drinks and tall stories with him while Patton was stationed with the 13th Calvary Regiment at Sierra Blanca. The advice proved sound, as Patton's bullets "broke the horse's hip and he fell on his rider." "Impelled by misplaced notions of chivalry," Patton wrote, "I did not fire on the Mexican who was down until he disentangled himself and rose to fire." Holmdahl had no such inhibitions, and as the man staggered to his feet, the scout shot from ten feet away, killing him.

The Mexican, whose horse had been gutted by Patton's first shot, was now up and running. Patton recalled, "I saw the man about 100 yards off, I shot three times at him with my rifle, four or five others fired also and he went down."[12] Meanwhile, the man who Patton had shot in the arm as he fled through the gate, wheeled his horse and rode back into the hacienda courtyard. Easing himself out of the saddle, dripping blood from his shattered arm, he ran into the ranch house and dashed to the back windows.

The Palace Hotel in Sierra Blanca, Texas. It was here that gunfighter Dave Allison told Patton, "When attacked by a man on horseback, shoot at the horse." — Courtesy Sierra Blanca Museum

Pulling himself through the window, he dropped to the ground, and firing his pistol with his good arm, he ran along a fence at right angles to the road. He was 300 yards away, when Patton spotted him and yelled to one of the troopers. The soldier calmly wrapped the sling of his Springfield about his arm, took a bead on the running man, and fired. The trooper shook his head and cursed as his first bullet missed. Aiming again, he fired and the man dropped. Turning to Patton, the trooper held up one finger and smiled. Patton later remembered, "It was remarkable how cool the men were during the fight."[13]

The wounded man, lying face down, lifted his head and with an effort raised his left hand in a gesture of surrender. But as Holmdahl strode toward him, the man's face twisted into a grimace of hate as he recognized the former Villista officer. With a final effort, he raised his pistol and fired.

Holmdahl, his sardonic grin never changing as the bullet sped past his ear, drew his pistol, and coldly shot his old saddle-mate, Colonel Julio Cárdenas, late commander of Villa's "Dorados," through the head. William Walker, one of the civilian drivers, later charged that Holmdahl was an adventurer who "killed for pleasure."[14] Cutthroat though he may have been, Julio Cárdenas died game. Aside from Patton's bullet that broke his right arm and Holmdahl's *coup de grace* to the head, the Villista officer had been hit with two other bullets that ripped through his lungs. The trooper's first shot did not miss after all.

Not lingering over the bloody body of the dead colonel, Patton made a quick survey of the scene. Believing Cárdenas might have as many as thirty men with him, he realized his troops were vulnerable to fire coming from the crenelated roof of the hacienda. Spotting a dead tree trunk lying against the wall of the house, he ordered two troopers to prop it against the side of the building.

Holstering his pistol and slinging his Springfield over his shoulder, Patton shinnied up the tree trunk and stepped out on the roof. As he did so, the roof collapsed under him, and he dropped down through the ceiling to his armpits.

Wiggling free, he perched on the shaky roof and scanned the deserted courtyard. Satisfied that there were no more Villistas lurking about, he climbed down, and, as his men gathered around, he ordered a room-to-room search of the ranch house and its outbuildings.

Patton ordered the three men and a boy, who throughout the fighting had ignored stray bullets whizzing by them, brought to him. Pushing the cow skinners before him, he used them as a movable shield as the Americans searched the labyrinth of rooms in the large *hacienda*. Finding the rooms locked, Patton used his revolver to shoot off the locks of each room before entering.

In one room they found, silent and tight-lipped, the wife, now the widow, of the Mexican colonel. She was holding his infant child in her arms. His mother stood beside her, staring defiantly at the intruders. In another room, they found two old women whimpering in a corner. A final search turned up a saber and a silver-mounted saddle, confiscated by Lt. Patton as trophies of war.

The troopers collected the weapons of the three dead Villistas and stowed them in the autos. Then they hoisted the bodies, one to a car, and spread-eagled them over the hoods, tying them down. Suddenly, Holmdahl gave a shout and pointed to a body of approximately forty horsemen heading their way at the gallop. Patton, putting discretion before valor, weighed the odds and ordered his Dodge caravan to drive off at full speed. After a few miles when his pursuers were lost from sight, they halted.

Patton pointed to the telegraph lines paralleling the road. "Cut 'em," he ordered. A trooper climbed a pole and severed the wires with a bayonet. As they drove through Rubio on the way to Pershing's headquarters, the young lieutenant did not want a reception committee waiting for him. As it was, the trio of Dodge cars got some hateful stares as they passed through the town.

Arriving at Pershing's headquarters about four o'clock that afternoon, they created a mild sensation. The little caravan bearing the bloody bodies stretched out on the hoods drove through the camp like a band of proud deer hunters displaying their kills. The two corpses, other than Julio Cárdenas, were soon identified as Captain Isadore Lopez, a Villista veteran, and Private Juan Garza. On a plaintive note, in Lopez's shirt pocket they found an unfinished letter to his sweetheart.

Patton, in an ebullient letter to his wife back in El Paso wrote,

> *The General* [Pershing] *has been very complimentary telling some officers that I did more in half a day than the 13th Cavalry did in a week...You are probably wondering if my conscience hurts me for killing a man. It does not. I feel about it just as I did when I got my swordfish, surprised at my luck.*[15]

It seemed the West Point lieutenant and the hardened Holmdahl had much in common.

American newspaper correspondents assigned to the punitive expedition wrote gory dispatches that made the young lieutenant a national hero. One correspondent, Frank B. Elser of the *New York*

Times, described the affair as "a fight that will go down as unique in the records of this expedition."[16] Patton, he pointed out, eschewed the regulation army Colt .45 automatic pistol and dealt out death with his personal ivory-handled six-shooter. But Patton ruefully reported to his wife that his fellow officers were teasing him for shooting the bandits rather than sabering them.

Ironically, Patton, one of the last of the romantic cavaliers charging on horseback with drawn sword, led one of the first American actions where cavalry dismounted from automobiles and fought on foot with rifle and pistol.

General Pershing, Patton told his wife, was jokingly referring to him as his "bandit." Other soldiers, after eyeing the Dodge autos with bodies dripping blood and gore, came up with another nickname. An old cavalry colonel, "Old Pants" Johnson, looking at the broken bodies of the Villistas remarked, "Look at the dirty bastards; look at the blood and guts on those dirty bastards."[17] And a legend and the nickname, "Blood and Guts," were born. Through it all, Holmdahl was said to have never stopped grinning.

At that time Holmdahl was still out on bail and was seeking a pardon for his federal conviction. He probably solicited the following letter from Patton:

> *Headquarters United States Troops*
> *Somewhere in Mexico*
> *May 20, 1916*
> *To Whom it may concern:*
> *This is to certify that Mr. E.L. Holmdahl, was the Government Scout with the U.S. Troops under my command in an engagement with Villa Bandits, at San Miguel Ranch, Chihuahua, Mexico, on May the 14th. I have highly recommended Scout Holmdahl, for his coolness, courage and efficiency while under fire, he personally killed General Julio Cárdenas, and Colonel Gildardo López, in a pistol duel. At that time Holmdahl fought in the open, without cover of any kind and shot with great accuracy and deliberation his action being that of a man at target practice.*
>
> *I also wish to recommend him to any brother officer, who may wish a man who is thoroughly familiar with Mexico and its people or in any position of trust, as he is most reliable, and a Good Soldier.*

(Sgd) Geo. Patton.
1st Lieut 10th, U.S. Cavalry.
A.D.C. General Pershing.[18]

On July 10, Holmdahl was off on another scouting expedition for Pershing, as attested by the following pass:

Commander of the Guard
Camp of Colonia Dublán
Please allow scout Holmdahl to pass the lines on Govt. business this date. July 10, 1916.

Respt.
W.W.Reed
Capt. 6th Cavalry
Asst. Chief Staff [19]

Holmdahl was probably on a mission seeking Villa's whereabouts, and he may have succeeded. One can only speculate, but two weeks later he was back at Pershing's headquarters at Colonia Dublán where he was discharged. This coincides with Holmdahl's claim to historian McGaw that after he traveled to El Paso, a Pershing aide offered him $100,000 to assassinate Villa. His orders read as follows:

Headquarters Punitive Expedition, U.S. Army.
In the Field, Mexico, July 24, 1916.
From: The C.O. Hdqrs. Det. Punitive Expedition, Dublán, Mex.
To: The Quartermaster at the Base, Columbus, N.M.
Subject: Guide Holmdahl
1. Guide E.L. Holmdahl has been discharged and will leave for the border on the first transportation. Please pay him for services as guide from July 1st 1916 to date of his arrival at Columbus, N.M., Rate of pay One Hundred and Fifty Dollars, ($150.00) per month. He is entitled to transportation to El Paso, Texas.

M.C. Shallenberger
1st. Lieut., 16th Inf., ADC
in charge of guides and scouts. [20]

Shallenberger was also the officer in charge of the expedition's intelligence operations.

Whether Holmdahl was discharged in anticipation that he would accept the assassination offer, and it was expedient for the army to sever all official connections with him, is not known. But by the end of July, Pershing was increasingly frustrated by the limitations put on his command by Washington, D.C. after Carranza, following the Parral fiasco, demanded the expedition be withdrawn from Mexico.

Pershing wanted Villa badly, and it seems he was not particular how he got him. Holmdahl's refusal to accept the mission to kill Villa may have angered Pershing, and he well might have refused to let Holmdahl re-enlist as a scout. Holmdahl disappears from view during the next four months, but in December while in El Paso, he received a curious telegram from the commanding officer of the Southern Department of the U.S. Army, based in San Antonio, Texas:

> *Received at 107 North Oregon Street, El Paso, Texas*
> *E.L. Holmdahl*
> *Your telegram twenty sixth offering your services is at hand. Your name has been given to intelligence officer who is keeping track of all persons like yourself who might be of great assistance in case of a general movement into Mexico Thanks for your offer.*
> *Frederick Funston*
> *Major General* [21]

The general probably saw a kindred soul in Holmdahl, who received the following telegram:

> *Dec 26 1916*
> *To: Scout Holmdahl*
> *Hdqts 606 Brigade*
> *You are ordered to scout out around the Mesquite and report best location for next meeting of the staff. Should you encounter any of the enemy shoot on sight.*
> *By order of Cmdg General Flynn*
> *Adj. GW* [last name undecipherable] [22]

By this time, the United States and Mexico were very close to war, as Carranza increased his demand for the immediate withdrawal of the punitive expedition. In mid-June, Pershing, fearing an attack by the Mexican army, received information that a large force was building up at the railroad town of Villa Ahumada, about eighty miles south of El Paso. If war broke out, those troops would be on his left flank rear and could threaten his supply route from Columbus. Pershing issued written orders to Captain Charles T. Boyd, an experienced West Point-trained cavalry officer, to reconnoiter the area and specifically to "avoid a fight." It was said, however, that Pershing had a secret private conversation with Boyd before he departed.

On June 20, while camped at an American-owned ranch, Boyd announced that on the following morning he would ride through the little town of Carrizal, heavily garrisoned by Carranza troops. When his two fellow officers and a civilian scout protested this would not only violate his written orders but would precipitate a battle, Boyd muttered something about "In the morning we will make history."

The next morning he attacked Carrizal, eight miles west of Villa Ahumada. Boyd dismounted his eighty cavalrymen and charged on foot across 300 yards of open ground into 400 veteran Carranza soldiers sheltered in an irrigation ditch. The Carranza forces were armed with Mauser rifles and two machine guns. Tactically, it was like meat fed into a grinder. Boyd and one other officer were killed, two dozen more were wounded, and twenty-three were captured. The rest fled west across the desert until picked up by U.S. patrols during the following days. It was the sole defeat suffered by the expedition and it came not at the hands of Villistas, but by troops loyal to the Federal government. When one studies this fiasco, one can conclude that either Boyd had taken matters into his own hands or that perhaps Pershing had whispered unofficial orders in his ear.[23]

Chafing at the restrictions placed on his command, Pershing may have wanted to create an incident that either would lead to war with Mexico or encourage Washington to lift the restraints and enable him to further pursue Villa. The truth will never be known. Although hotheads on both sides of the border clamored for war, cooler heads prevailed, and negotiations for the expedition's withdrawal were conducted by the American and Mexican governments. As a result, Pershing pulled in his patrols and drilled his troops in camp.

As Pershing's cavalry pushed deeper into Mexico during March and April of 1916, Villa and most of his band scattered and retreated into southern Chihuahua. Reconstituting his forces there, Villa was able to raid Chihuahua City and spread havoc as far south as Durango. But while he was still a regional menace, his ability to seize power in Mexico City was broken.

The 300 yards of open ground across which the 10th Cavalry "Buffalo Soldiers" charged into machine-gun fire from the trees. — Douglas V. Meed Collection

In later years he became more of a nuisance than a threat to the central government. By 1920 he had alienated many in his Chihuahua strongholds, and President Adolfo de la Huerta was able to purchase his abdication from politics with a generous gift of land and money.

Many considered Pershing's expedition a failure because he did not kill or capture Villa; but Pershing was not a sheriff and the U.S. Army was not a posse.

Driving Villa and his men far from the American border successfully protected American towns and properties from further raids. This, in fact, was the major strategic purpose of the expedition. Also, breaking up Villa's main forces early in 1916, enabled the Carranza government to consolidate its power over Mexico and begin to introduce political and economic stability.

It can be argued that the worst possible outcome of the expedition was to kill or capture the elusive Pancho Villa. To have killed him in action would to have made him a martyr with the United States being the chief villain. To capture him would have been worse. He would have been taken to New Mexico, tried for murder, convicted, and subsequently hanged in the back yard of the Deming, New Mexico jail, along with the other Columbus raiders. His execution would have been a lasting symbol to rally hatred against all Americans.

In preparation for war against Germany, President Woodrow Wilson ordered the expedition to return to the U.S. In January 1917, the expedition began to pull back across the border. By February 1, all American troops had been withdrawn from Mexico.[24]

On April 6, 1917, the United States declared war on Germany and Austria-Hungary. On May 10, Pershing was summoned to the office of General Hugh L. Scott, the army chief of staff, and told that he would command the American troops who would be sent to France.[25]

Meanwhile, Holmdahl was eager to get into the Great War. But as a convicted felon no longer working for the army, he was in danger of having his bail canceled and being forced to serve out his eighteen-month sentence in a federal prison. Using all the influence he had built up over the years, he increased his campaign to get a presidential pardon.

During World War I, Holmdahl was commissioned a captain in the U.S. Army Engineers. He fought alongside the British during the last German offensive in 1918.
— Holmdahl Papers

≫ 12 ≪
With the A.E.F.

Over there, over there,
Send the word, send the word to beware
The Yanks are coming.
— George M. Cohan

During his many years on the border, Holmdahl managed to be known and liked by many influential people in the American Southwest. In February 1917, he began an intensive letter-writing campaign to secure a pardon, and one of his first letters was to the attorney general of the United States explaining his actions. It was an ingenuous document, but not quite in conformity with the trial evidence. Also, in the letter Holmdahl gave another indication that he had acted as an intelligence agent for the Department of Justice. Perhaps that is why in the coming months they treated him so leniently.

Washington, D.C.,
February 1, 1917.

The Honorable
The Attorney General,
Washington.
Sir:
In reference to my application for pardon, following my conviction in the United States District Court for the Western District of Texas, on the

charge of having engaged in a conspiracy to violate the neutrality laws of the United States, I beg to respectfully state that at the time of the alleged offense was committed, I entertained no thought whatsoever of violating said laws or any statute of the United States. The facts in my case are as follows:

I was employed by General Benjamin Hill, commanding the Carranza forces in Chihuahua and Sonora, to gather confidential information for him along the border, the latter part of 1914. At that time General Hill indicated to me his desire that I should come over to Chihuahua and take charge of certain of his troops, with a view of attacking Juárez. Nothing was said by him to me that these troops were to be recruited in the United States, and I naturally assumed that they would be a part of his own command. Meanwhile, the Carranza consul in El Paso informed me that he had purchased a carload of ammunition and equipment, and there being no embargo he was to dispatch the car to Douglas, Arizona, for the delivery of its contents across the border. After the car had left, the consul received information that the Villistas had planned to steal the car at Mimbrios twelve miles from Columbus, and the consul sent me there in an automobile to ascertain whether the theft had been accomplished. I did not find the car and learned that the American troops had escorted the car through to Douglas, where it was delivered to the Carranzista authorities at Agua Prieta. Prior to this incident, I met a Colonel Valle, of General Hill's staff, at El Paso, who desired to send some code telegrams relating to this matter to the Carranza representative in Douglas, viz: Francisco S. Elías, but the telegraph office had refused to accept them because Valle desired to dispatch them collect. He inquired of me if I could send them for him, which I did. I did not know the contents of these messages, nor do I know them yet. It appeared from the testimony given in my case that the Mexican consul, Jorge Orozco, and Victor L. Ochoa and José Orozco, all of El Paso, had engaged the services of some Mexicans to cross the border and join General Hill's forces. The testimony showed that I had no connection with these men, and never saw any of them. This was corroborated by the testimony of the two Orozcos and Ochoa, and there really was nothing connecting me directly with the enterprise. There was no testimony adduced showing any real guilt on my part or showing that I had really conspired with any one to violate the law.

Indeed, there was no proof that I intended to accept General Hill's invitation to join him in Chihuahua, and I am sure that a mere cursory review of the evidence will substantiate my assertion.

I have never concealed anything from the United States authorities on the border; have always been frank in dealing with them, and have, as a matter of fact, from time to time given them information of the most valuable character, as will be borne out by General George Bell, U.S.A.; Major General Pershing, U.S.A.; District Attorney Crawford, and Special Agent Stone of the Department of Justice. I beg that the foregoing may be taken into full consideration in determining my application for Executive clemency.

Respectfully,
(Sgd.) E. L. Holmdahl. [1]

To better press his case, Holmdahl temporarily moved to Washington D.C. while an obliging Justice Department continued to extend his bail and allow him to travel.[2] On February 17, Colonel J.A. Ryan wrote, "I . . . am always ready to testify to your good service with me and the Expedition in Mexico . . . Keep in touch with me as I may need you if this matter breaks out either in Mexico or Cuba."[3] On February 22, Holmdahl replied, "When war, or intervention short of war, comes, either in Mexico or Cuba, or elsewhere, I certainly hope to be among those to follow you."[4]

Earlier, Holmdahl had asked Jeff McLemore, a member of the U.S. House of Representatives from Texas, to intercede for him and on March 19 the Congressman sent him copies of two letters written by his old boss.[5]

Headquarters.
Southern Department
Fort Sam Houston, Texas
March 11, 1917.
Honorable Jeff McLemore
U.S. Representative
Washington D.C.

> *My Dear Mr McLemore:*
>
> *With reference to your letter of February 14th, setting forth the case of E.L. Holmdahl, I have fully investigated the case and am now prepared to change my opinion.*
>
> *It appears from the reports of the investigation which I instituted that Holmdahl, was "more sinned against than sinning", and insofar as his conduct on that occasion is concerned, I am now willing to make recommendation for clemency.*
>
> *This has no reference, of course, to his services as a scout during the Punitive Expedition. While he did some good work there, he did other things which were disapproved by me.*
> *Sincerely Yours,*
> *(Sgd) Major General John J. Pershing*
> *U.S. Army Commanding.*[6]

And then Pershing wrote to the Attorney General of the Army:

> *I desire to recommend executive clemency in the case of E.L. Holmdahl, charged with violation of the neutrality laws on the Mexican border and sentenced to eighteen months imprisonment. An inquiry into the facts of this case leads me to believe that Holmdahl should not be held to more than a technical violation of the neutrality laws.*
>
> *Moreover, as a scout with the Punitive Expedition into Mexico, he performed service which should entitle him to the Government's consideration. I trust that you will give this request such consideration as you may deem advisable.*[7]

Pershing did not elaborate on his comment that "he did other things which were disapproved by me." There are a number of possibilities, including Holmdahl's failure to accept an assignment to assassinate Villa. Pershing was possibly annoyed at the spectacle of the three dead Villistas draped over the hoods of Lt. Patton's automobiles. Or he might have objected to a newspaper story that claimed Holmdahl had hunted down and gunned down three Villista officers responsible for murdering seventeen American mining engineers who were taken off a Mexican train on January 9, 1916, and summarily shot.[8]

General Hugh Scott, who back in the Philippine days signed one of Holmdahl's fitness reports and was undoubtedly aware of his intelligence reporting to the U.S. government, probably also interceded for him. On March 16, Holmdahl wrote the U.S. Attorney in El Paso asking for a sixty-day "respite" before reporting to Leavenworth Federal Prison. The respite was granted.[9] In the meantime the U.S. declared war on Germany and Austria-Hungary.

On April 21, Mayor Tom Lea of El Paso wrote to the U.S. Attorney General requesting executive clemency and a pardon for his old friend. "Although he is a soldier of fortune," Lea wrote, "his reputation is most excellent." The mayor added,

> *Holmdahl has been of very great help and benefit to the Police Department in many ways. He asked me to speak especially with reference to his arrest in El Paso some year and a half ago, in which he was accused of a misdemeanor . . . I know that Mr. Holmdahl acted properly on that occasion, and resented an insult from an anti-American. He was released without bond and acquitted on proper bail.*[10]

There was a note of desperation in a letter Holmdahl sent to Frank Polk of the U.S. Department of State on April 30. He wrote:

> *. . . my application for executive clemency . . . only awaits your recommendation.*
>
> *I have now exhausted both my savings and my credit, and am two thousand miles from home, without any means whatsoever . . . begging your kind and early consideration.*[11]

Funds exhausted, Holmdahl returned to El Paso, where he began to blossom out as a raconteur of border adventures. Joe Goodell, the owner of El Paso's Sheldon Hotel, described Holmdahl as, "Sporting a black diamond ring on his finger, [he] wore flamboyant clothes and attracted attention wherever he went." Goodell said he provided the famous mercenary, gunrunner, and scout with a free room, "Simply in repayment for his drawing power and influence over prospective patrons."[12]

Holmdahl spoke articulately in a clear voice, and when he recounted some of his wilder adventures, he laughed. Sometimes, however, the laugh was only on his lips; his eyes stayed cool. He gave the impression he could be a good friend but a terrible enemy. Above all, he was a soldier.

Adding to Holmdahl's restlessness was the news that his old comrades were also back in harness. Tracy Richardson was fighting in France as a captain in Princess Patricia's Own Light Infantry, one of the elite regiments of the Canadian Army.[13] Sam Dreben, who was to win renown as one of the most decorated American soldiers in World War I, was a first sergeant with the 36th "Texas" Division.[14]

As the American Expeditionary Force began to mobilize for overseas duty in France, another old soldier from border days swung into action. Major Sam Robertson, commanding officer of the Sixth Reserve Regiment of U.S. Engineers, telegraphed Congressman McLemore:

> *Won't you see the Attorney General and endeavor to get Holmdahl pardoned at once. Regiment needs his services badly and he will be more valuable to his country in France than in prison.*[15]

Robertson sent an identical telegram to Frank Polk at the State Department.[16]

It took another month, but on July 13, 1917, Holmdahl, now back in Washington D.C., was granted a full and unconditional pardon by President Woodrow Wilson.[17] He immediately went to Washington Barracks and enlisted as a private soldier in the 6th Engineer Regiment. When he took the physical, however, he was disappointed.

The army medical corps physician, Lt. H.L. Taylor, found Holmdahl unfit for service by reason of pain and limited flexibility in the scout's right knee. This was due to shrapnel wounds received in 1910.[18]

In a later army physical, Holmdahl was reported to have suffered wounds in 1911-12-13-15-18. They included, "'so far as known', shrapnel in right knee two, right shoulder, bullet two in breast, and stomach."[19]

Adjutant General G.W. Read fired back a message to the medical officer, stating, "Enlist that man if he has only one leg."[20] General Read followed up with a memo to the War Department Adjutant General Files: "The Secretary of War authorizes the enlistment of Emil L. Holmdahl, for the 6th Regiment, Engineers, National Army, waiving the defects reported."[21] After nineteen years, Holmdahl was again a private in the United States Army.

On July 21, Holmdahl wrote Senator Morris Sheppard, who had helped secure his pardon, stating, "I will do my very best in the Army to vindicate [your] confidence . . . And at the very least I can do is to sacrifice my life for a worthy cause."[22] On that same date, in a more personal vein, he wrote Representative Jeff McLemore, who, he said, had been "a soldier of fortune yourself," thanking him for his help. A bond of friendship had sprung up between Holmdahl and McLemore.

Holmdahl wrote,

> *I shall never forget, and at any time that I can be of service to you or any of your friends, I will fight to the last ditch . . . Now that I am into it I shall try to live down the reputation given me on the border, or live up to it and fight harder than ever.*[23]

Always friendly with Department of Justice officials, Holmdahl wrote James A. Finch, an attorney involved with his pardon applications, and received back a semi-humorous reply stating, "By the way, you are the only one so far who has ever gotten away with the stunt you did. Please remember you have the good wishes of all of us . . ."[24]

Holmdahl reported to the 6th Engineers, later designated as 16th Engineers. Several days after his enlistment he was promoted to first sergeant. Two weeks later he, with his regiment, were on a troop ship headed from France. Within weeks, Holmdahl was promoted to second lieutenant and then quickly to first lieutenant.[25] Even before combat units of the U.S. army went into action, Holmdahl and his engineers were fighting as infantry alongside the British.

For the raw Americans and even for a veteran like Holmdahl, the Western Front was a horror almost inconceivable to civilized men. British Prime Minister David Lloyd George summed it up when he wrote:

> *When I read of the conditions under which they fight, I marvel that the delicate and sensitive instrument of the human nerves and the human mind can endure them without derangement.*
>
> *... troops are called upon to live for days and nights in morasses under ceaseless thunderbolts from a powerful artillery, and then march into battle through an engulfing quagmire under a hailstorm of machine gun fire.*[26]

The prime minister put it mildly. In the trenches of France there was also the stench from rotting flesh and gases escaping from bloated bodies left in "no man's land." There was cold food on freezing days; toes turned black from soaking in putrid water; and always the haunting fear of miserable and sudden death from snipers, shellfire, mines, or poison gas. If there ever was glamor or glory in war, it ended in rat-infested trenches on the Western Front.

U.S. engineering units began arriving in France in early August. They were the first U.S. troops to be sent to the front in response to the urgent request from the British. The British forces desperately needed trained railway personnel to construct tracks and aid in loading British tanks in preparation for a fall offensive. Holmdahl and his engineer unit were immediately assigned to aid the tank corps.

In October 1917, communists overthrew the moderate Russian reform government that had succeeded the Tsarist regime. The announced policy of the Bolsheviks was to seek an immediate peace with Germany. This, the Allies believed, would result in a massive shifting of German troops to the Western Front.

General Sir Douglas Haig, the British commander-in-chief, planned to smash Germany's western armies before they could be reinforced. At dawn on November 20, he launched a massive surprise attack on the German trenches without the usual artillery bombardment. Then, for the first time, Winston Churchill's secret weapon, called the "tank," was used in large numbers.

The early British tanks weighed approximately thirty tons, were more than twenty-six feet long and were armed with two six-pounder cannon, four machine guns, and manned by a crew of six. They had less than one-half-inch of armor plating, sufficient to ward off rifle and machine-gun bullets, but they were vulnerable to artillery. The "landships" could plough through barbed wire and cross narrow trenches, while plugging along at a top speed of 4.7 miles per hour. Breakdowns were frequent and they had a range of approximately twenty miles before they ran out of fuel.

To give them any fighting range at all, they had to be transported by rail close to the front lines. Thus, the need for engineers to construct roadways and lay train tracks. By mid-August elements of three railway engineer regiments, including Lieutenant Holmdahl, were busy building standard-gauge and light-railway lines for the British military.

Great secrecy was maintained in the operation, with the tanks parked in a hidden valley until just before the attack. Then they were loaded on flatcars and hauled to the front lines, where they were unloaded by the engineers. On November 20, more than 400 of the lumbering behemoths surged across "no-man's land." They smashed through the German barbed-wire entanglements, in some areas fifty yards deep, and machine-gunned and shelled the enemy trenches. They made a breech in the Hindenburg Line four miles wide, captured 10,000 prisoners, 200 pieces of artillery, and penetrated five miles deep into German defenses.

When the news of the breakthrough was first reported, all the church bells in London rang. But not for long. It was all for naught, because sufficient infantry reserves were not available to pour into the breech and exploit the breakthrough. The German army quickly recovered from the initial shock of the attack and reinforced the broken sector. On November 30, they launched a counterattack against the exhausted British and recaptured all the ground lost, driving three miles deep into the British front.

The American engineers dropped their shovels, picked up their rifles, and fought alongside the British infantry. Under heavy shellfire, they suffered many casualties, until their British allies finally stabilized a defense line.[27] The British and German armies each suffered more than a quarter of a million casualties for no discernable change in the stagnant Western Front. The American engineers continued repair and maintenance work on the British railway system, until in the early spring of 1918, when all hell broke loose on the British sector of the Western Front.

By March, the German army, after concluding a peace with Russia, moved fifty-two divisions from the now-peaceful eastern front to the west. The chief of the general staff, General Erich Ludendorff, and the commander of the German army, Field Marshall Paul von Hindenburg, decided they must strike the British and French quickly before millions of American soldiers began arriving. On March 21, under cover of a dense fog, they launched a massive attack, hurling their entire reinforced armies against the Allied lines. The British army, with most of its reserves used up after four years of fighting, fell back in exhaustion.

On that date, Holmdahl and the 6th Engineers were quartered at the town of Doingt, where they came under "very severe artillery fire." The regiment was ordered to retreat to Chaulnes, the site of a major supply dump. When they reached the dump on the morning of March 23, they were ordered to destroy all supplies stored there, as the German advance was about to overrun them. Blowing up everything but their backpacks and the trucks to transport them, the engineers began a long retreat, until on March 27, they reached a wooded area named Bois de Taillaux.[28]

There, they dug trenches alongside British infantry units under the command of British Brigadier Sandeman Carey. Nicknamed "Carey's chickens," as they were under his wing, the engineers prepared for a "last stand" as the German army came close to breaking the entire Allied line and winning the war.

By telephone, messenger, flag signals, and military police patrols, Carey rounded up a motley group of non-combat units required to fight like Guardsmen. There were labor battalions of middle aged men, electricians, truck drivers, cooks and bakers, stragglers, plus fifty stray cavalrymen. And, there was Lieutenant Holmdahl and his railroad engineers.[29]

Some of the men were armed with rifles; others only had pistols which were virtually useless. Fortunately, there was a British machine-gun school several miles behind the lines. These guns were brought up to the "chickens," although few of the men knew how to operate them. Enter Holmdahl. With a decade and more of using rapid-fire weapons in combat, he blossomed as an instructor for both British and American troops.

It was in that moment of desperation that General Sir Douglas Haig made his famous declaration:

> *Many among us now are tired. To those I would say that victory will belong to the side which holds out the longest . . . There is no other course open to us but to fight it out! Every position must be held to the last man; there must be no retirement. With our backs to the wall, and believing in the justice of our cause, each one of us must fight on to the end . . .*[30]

For six days the "chickens" poured a scythe-like stream of bullets at wave after wave of German troops attacking their makeshift trenches. They held against "continuous assault" until April 3, when they were relieved by fresh British combat units. Their tough defense had prevented the German artillery from pushing on to a point where they could have shelled Amiens, a vital railroad center which kept the entire British army supplied with food and munitions. A U.S. army report stated "their sudden desperate stand . . . robbed the Germans of complete victory."[31]

General Haig, a man of few compliments, later wrote:

> *I am glad to acknowledge the ready manner in which American engineer units have been placed at my disposal . . . and the great value of the assistance they have rendered . . . British and American troops have fought shoulder to shoulder in the same trenches, and have shared together in the satisfaction of beating off German attacks.*[32]

As the German attacks began to lessen, Holmdahl's unit was assigned to work with an Australian corps rebuilding bridges until June 10, 1918, when they were assigned as divisional engineers for the Third American Infantry Division. On July 14, while Holmdahl's Company F was constructing defense works on the Third Division front near Chateau Thierry, the Americans suffered a heavy artillery bombardment, followed by a strong German attack on the morning of the 15th.

The attack was beaten off, and a week later the Americans were ready to advance. Holmdahl's company, in preparation for the Third Division assault on July 21, constructed two footbridges across the Marne River. The following day, after the Americans had attacked and advanced, Holmdahl and his men spanned the river with a pontoon bridge built from captured German equipment.[33] On July 30, 1918, Holmdahl was promoted to Captain.[34]

After the bridges were built and the American advance continued, the engineers were "engaged as infantry support." They continued fighting until on August 10, when they were relieved and moved to a rear area where they rested and refitted. Holmdahl was assigned to the Engineer School at Longre, France to train newly arriving American troops. On November 23, after the Armistice ended World War I, he crossed the Atlantic and was assigned to Camp Leach, Washington D.C.[35]

Holmdahl's World War I records were lost during a fire at the Federal Documents Depository in St. Louis, along with much of the documentary evidence of his World War I service. In an interview given to the *Los Angeles Times* in 1967, Owen W. Miller of Hermosa Beach, California, who served with him in the 6th Engineer Regiment in France, recalled:

> *He was quite a mystery man. The rest of us were construction men but he was obviously a soldier of great experience. Later he transferred from the engineers to a combat regiment.*[36]

According to his nephew, Gordon Holmdahl of Dublin, California, "Uncle Emil returned from France with a stomach full of shrapnel."[37] After more than one year on the firing line, Holmdahl had more than fulfilled the vows he made to those who had arranged his pardon. He promised to serve his country well during the war, and he did.

Emil Holmdahl in Mexico in the 1920s. — Holmdahl Papers

13
Drifting

Better to see your cheek grown sallow
And your hair grown gray, so soon, so soon,
Than to forget to follow, follow
After the sound of a silver horn.

— Elinor Wylie

Captain Holmdahl remained at Camp Leach until early July, 1919, when he was assigned to a government surplus sales department. Possibly because of his language skills and his knowledge of railway equipment, he sailed to France on July 17. In addition to disposing of railroad tracks, engines, flatcars, and bridging equipment left behind after the departure of the A.E.F., Holmdahl was authorized to sell various components of artillery shells and small arms ammunition piled up in warehouses both in France and the United States.

After a month in Paris, Holmdahl traveled to Madrid to arrange sales of surplus locomotives and railroad supplies to the Spanish government.[1] In September, his work completed, he returned to the states, where he commanded a desk at the Purchase, Storage and Traffic Division at the Office of the Director of Sales for the War Department. It was a job that had dubious appeal to a man of Holmdahl's temperament. He submitted his resignation effective June 1, 1920, and was soon off again on further Mexican adventures.[2]

After his discharge from the army, Holmdahl, finding himself adrift, headed south to El Paso, where he became involved in a shooting scrape. As he tells the story, he and a friend were searching for a place in which to enjoy a Sunday lunch, when they were attracted by guitar music coming from a restaurant.

After entering the restaurant, a deputy constable named W.C. Boyle burst through the door and screamed, "Put up your hands!" Holmdahl spun around as Boyle fired a shot from his revolver which struck a man named W.E. Davis, apparently an innocent bystander. Holmdahl shouted, "Don't do that," grabbed the constable's hand and twisted the pistol out of his grasp.

The badly wounded Davis, with a bullet lodged in his side, was taken to a nearby hospital by Holmdahl and Boyle. The Constable suddenly disappeared when Holmdahl called the police, and Boyle was later arrested and charged with assault to kill. The constable later stated he had intended to arrest a felon in the restaurant, but his revolver went off and hit Davis by accident.

Police officials said Boyle had a habit of staging "wild west stunts" and this one had gone badly awry. Davis survived; Boyle went to jail nursing a sprained thumb as a result of Holmdahl's twisting the gun out of his hand. Holmdahl and friend returned to their deferred lunch to the strains of guitar music. Even for Holmdahl, it was not an uneventful Sunday luncheon.[3]

According to a newspaper account written by William C. Stewart, a *Los Angeles Times* staff writer who knew Holmdahl, at about this time, the veteran soldier launched a treasure hunting expedition into Mexico. According to Stewart,

> Holmdahl began thinking of Villa's treasure buried by Urbina and he set out to try and find it. He was accompanied by the reformed train robber and later movie actor, Al Jennings, who had transferred his activities from Oklahoma to Los Angeles.

Al Jennings billed himself as "The Last of the Great Train Robbers" in his religious revival speeches. More realistically, he should have been billed the most inept of the great train robbers. As the leader of the Jennings Gang, he fumbled a series of Oklahoma train and bank robberies from 1897 to 1898 before he and his band were captured and thrown into the territorial prison.

At one time a county attorney in El Reno, Oklahoma, Jennings was addicted to the bottle and the gaming tables. He often told fictional accounts of his facing down notorious gunmen in *mano-a-mano* confrontations. Temple Houston, the quick-shooting son of Sam Houston, killed his brother Ed Jennings and wounded another brother in a barroom shootout while feuding with the Jennings clan.

Along with his two remaining brothers and another pair of misfits, Jennings decided to make a reputation by robbing trains and banks. In their first "Great Train Robbery," they leaped aboard a Santa Fe train that was stopped for water, held up the engineer and broke into the express car. This escapade netted only a few hundred dollars, so two weeks later they attempted another train robbery near Muskogee, Oklahoma Territory.

They picked a spot where the railroad stacked spare road ties and grunting and sweating under a hot sun, they piled the ties across the track. Waving their guns ferociously, they signaled an approaching train to stop. The engineer, however, pushed his throttle wide open. At great speed the cowcatcher slammed into the ties, sending them flying into the air and nearly decapitating several of the Jennings gang. As the train chugged off to Muskogee, the outflung arm of the engineer signaled his opinion of Al Jennings and his band.

After several more fiascos, they robbed a Wells Fargo office, stealing several thousand dollars. The brothers Jennings fled to Galveston and boarded a tramp steamer bound for Honduras where they met a drunk named William Sydney Porter. Porter, who was later known as the great short-story author, O. Henry, was hiding out from a bank embezzlement charge in Austin, Texas. They probably met Lee Christmas and some of his hard cases and decided Honduras was not for them. After returning to Oklahoma, on October 1, 1897, they flagged down a Rock Island train a few miles from Chickasha, Oklahoma Territory.

Holding the crew at gunpoint, they broke into the baggage car, which contained two large safes. Planting four sticks of dynamite alongside them, they lit the fuses. There followed a mighty explosion which blew the baggage car to smithereens. The safes, however, remained intact and invulnerable. Frustrated, the bandits went through the passenger cars at gunpoint stealing watches and money from the passengers.

A few months later, a lone deputy located the gang, arrested them, and brought them to the jail at Checotah. Following his trial, Al Jennings was sentenced to life imprisonment at the Federal Penitentiary at Columbus, Ohio. There he was reunited with Porter, who had given himself up.

After five years, Jennings was pardoned. He wrote several books grossly exaggerating his career as a robber and gunman, and then began a more successful stint as a evangelical preacher and lecturer. He also played a number of small parts in Hollywood Westerns.[4]

How he met Holmdahl is not known, but it was, indeed, a strange pairing. About the expedition into Durango, Stewart wrote, "They collected a motley group of Mexicans for a bodyguard and set out for Urbina's old ranch. Before they could reach the spot a group of Mexican laborers stumbled on the fabled cache." Then the local police invited the treasure hunters to leave, pronto.[5]

With Jennings' luck, Holmdahl was probably lucky to get out of Mexico alive, much less find the buried gold of Tomás Urbina. The buried treasure business was to crop up on a number of occasions in later years. Did Holmdahl really believe he could find Villa's or Urbina's treasure? Was he running a confidence game with Jennings? Or was he using the "treasure hunt" as a cover for other, more secretive, activities? The answer is anyone's guess.

Perhaps some inferences as to Holmdahl's activities can be gleaned from the newspaper clippings he kept. One such story found in his effects was from the *San Francisco Examiner*, headlined,

> *Gun-running Plot Exposed. Adventurer Blamed in Conspiracy. Federal Officials Seek 'Master Mind' Attempting to Send Munitions to Insurrectos.*

The story asserted federal police officers in San Francisco and Los Angeles were "close on the trail" of the man who was behind a plot to smuggle contraband rifles and ammunition to Central American rebels. According to police, the guns were received at San

Francisco, trucked to Los Angeles, and then loaded onto fishing boats. The newspaper reported,

While a dozen or more men may be involved in the asserted plot, most of them are merely paid employees. The Federal officers here are said to be primarily concerned with capturing an American adventurer who is believed to have engineered the purchase of the contraband weapons.[6]

Among other adventures, Holmdahl cruised Mexican and Latin American Pacific coastal waters in a large yacht named *Paxinosa*. The craft, pictured in a Los Angeles newspaper, was described as a two-masted ketch of twenty-one tons equipped with an auxiliary gasoline engine. The story naming the owner and guests listed one "E.L. Humdahl [*sic*.] formerly a colonel of the Maderista forces in Mexico" as a passenger. The story headlined, "Craft Paxinosa Veers Southward For Cruise In Deep Waters," and reported they were "completely equipped and provisioned for deep sea cruising . . . around Lower California and west coast ports and other places of interest." [7]

It is, of course, possible that Holmdahl was relaxing on a pleasure cruise in southern waters; or perhaps he was reconnoitering. Perhaps the *Examiner* story did not pertain to Holmdahl. It is curious, however, that he would keep such a newspaper story for more than forty years.

The Urbina treasure fiasco apparently didn't satiate Al Jennings' desire to associate with Holmdahl. In early February 1926, the two and a Los Angeles businessman, Fred T. Hughes, bought 32,400 acres of land on a *hacienda* named "Corral de Piedra," ("Rock Corral").[8] Located at Rosario, sixty miles south of Parral, it was aptly named. The land was supposed to contain two placer mines and one hard-rock mine, with deposits of gold, silver, and copper.

Jennings, from his revival tent in Lawton, Oklahoma, in a newspaper interview, piously announced he was quitting the pulpit and that he "would no longer pilot sinners down the sawdust trail to God." After a failed campaign for the Democratic Party's nomination for governor of Oklahoma (he finished third), Jennings went into California real estate. He said he told the Lawton police chief,

"My conscience is clear . . . I told the chief the truth about California real estate . . . I've quit highway robbery jobs."[9]

It is unclear how the mine worked out, but apparently the partnership was soon dissolved. It was only a few weeks later that Holmdahl was back in Durango on an adventure that nearly cost him his life.

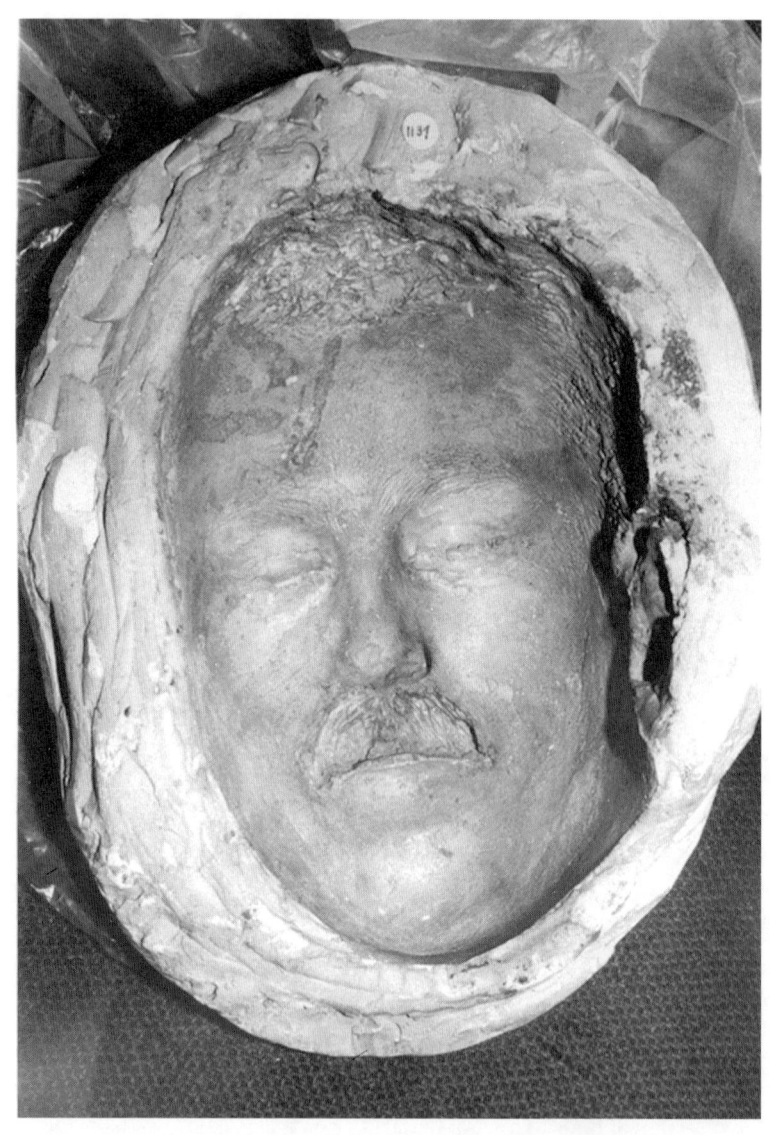

This death mask of Pancho Villa was made shortly after he was ambushed and shot to death in Parral, Mexico, in July, 1923. — Courtesy of the Radford School

❧ 14 ❧
La Cabeza de Pancho Villa

Such a strange infamy
Would never have been conceived
If gold had not been valued
Higher than manly honor.
The Yankees were not able
To defeat him in a fair fight
So they cut off his head.
  — *Corrido de la Decapitación de Pancho Villa*

In February 1926, Holmdahl was in the Durango mountains not far from Parral, searching for gold bars with a cousin of Luz Corral's, one of Pancho Villa's many wives. Probably he was still looking for either Villa's purported buried treasure or for more gold hidden by Thomas Urbina. He claimed that he and his companion, Alberto Corral, found the hidden treasure in a cave on the side of a cliff. With the help of two Indians, they lowered the gold 500 feet to level ground.

As they were loading the twelve-kilogram gold bars into their automobile, they were accosted by bandits. After facing down the bandits, who were apparently not very determined, or probably figured they might wind up suddenly very dead at the hands of this hard-eyed *gringo*, Holmdahl said he and Alberto drove to Parral. As they parked their car in front of their hotel, they were suddenly surrounded by police with drawn guns who arrested them. Within minutes they were rudely flung into a cell in the Parral jail.

Soon, Holmdahl related, a crowd of more than 2,000 gathered in front of the jail screaming for their blood. "I thought we were going to be lynched," he recalled. After passing a fearful afternoon and night, the "bewildered" pair were led out into the jail's courtyard at daybreak and marched to a bullet-pocked wall.

As they stared at a firing squad, they were accused of mutilating the body of Pancho Villa. They were told they must confess or be shot on the spot. Holmdahl said, at the time, he thought the whole episode was a plot to steal their gold, but he replied he didn't know what they were talking about and stoutly maintained their innocence. They were only poor prospectors, he maintained. Surprisingly, they were led back to their cells, but only for the moment.[1] Their arrest was the beginning of the mystery of the missing head of Pancho Villa.[2]

Along the Mexican border, legends about Pancho Villa are as prevalent as cactus in the desert. There are stories of gold buried in the mountains, and bloody tales of murder, betrayal, and tragedy. But the most mysterious of all are the tales which ask the question: "Who cut off the head of Pancho Villa?"

It is not surprising that most of the prime suspects were Americans. They were part of that wild bunch of adventurers who served as mercenaries with one or another of the many warlords who fought it out during the ten years of the Mexican revolution. Many claim to know who did the horrendous deed, but all their stories name different culprits. Some versions are burdened with facts while others are pure fancy.

For years the Mexican government had been fighting Villa in bloody battles ranging from the American border to deep into the interior of their battered country. In July 1920, a weary government decided it would be cheaper in both gold and lives to buy off Villa. They recognized that it was futile to attempt to track down and fight the elusive guerrilla in his many hideouts in the mountains of northern Mexico. They offered him an estate of 25,000 acres of good land in the state of Durango, 500,000 pesos and a paid escort of fifty men, in exchange for his solemn vow to never again take up arms against the government. Villa, who was momentarily tired of fighting, accepted, and for more than two years he kept his promise.

Villa took a new wife, although he had two other wives to whom he was "legally" married, and he and his henchmen spent their time raising cattle and improving the land on his huge estate. For recreation they despoiled a few local women, cheated more than a few local merchants, and raised hell in the bordellos of the nearby town of Parral. But in early 1923, rumors began to circulate that Villa, then a vigorous forty-six years of age, was thinking of making a political comeback.

The government in Mexico City panicked. On the sly, they contracted with men who hated Villa—and there were many—to assassinate the guerrilla leader. On Saturday evening, July 19, 1923, Villa loaded up his Dodge touring car with six of his *pistoleros*. Depending on which version you choose, they either drove to their favorite house of ill repute in Parral or to the christening of a friend's child. There they spent the night carousing and whooping it up.

Sunday morning, his bodyguards were hung over and still groggy from too many "cucarachas" (marijuana cigarettes), too many drinks of fiery tequila, or too many ladies of the evening. Villa's companions staggered out into the bright sunlight and piled into the Dodge, while Villa, who was not a drinker, insisted he drive the car.

On the outskirts of Parral, Villa slowed the car for a sharp turn. Suddenly, a blast of rifle fire ripped into the bodies of the bleary-eyed men. Villa, shot twice in the head and eleven times in the body, died instantly, as did most of his men. The bandit chief was buried with little ceremony in a quiet cemetery in Parral, where he lay quietly for two and a half years.

At six o'clock on the morning of February 6, 1926, Juan Amparan, the caretaker of the Pantheon Cemetery was making his early morning rounds when suddenly he let out a strangled gasp. By the opened grave of the mighty Pancho Villa, next to a shattered casket, lay his decomposing body. It was missing its head.

Amparan raced to the Municipal Hall of Parral and screamed for the mayor. Soon the gravesite was swarming with policemen and city officials. By the open grave, there was a large tequila bottle that gave off a sharp antiseptic odor and wads of cotton, one of which was soaked in blood. Odd, a policeman noted, a decomposing corpse, dead for almost three years, doesn't bleed.

A curious crowd gathered at the scene, and soon the police were screaming curses and hitting those who were pulling at the tattered clothes of the corpse to gain a souvenir of the famous general. Pieces of decaying flesh that came off with the shreds of cloth torn away by the crowd lent a touch of the macabre. Placing a guard around the grave, the officials pondered, "Who would do such a thing"?

The obvious suspect was Jesús Salas Barraza, a local politician and the admitted leader of the band who assassinated Villa and his bodyguards. It was said Barraza hated Villa because, he claimed, "the Hyena" raped his little sister who later died in childbirth.

Following his confession of Villa's murder, Barraza served less than a year in a local prison before an "understanding" governor of Chihuahua pardoned him. He was in Parral on the night of the beheading; however, he had an ironclad alibi.

Rumors were soon spreading like wildfire across the country. One story related that Villa, while on the run from Pershing's troops, buried vast amounts of silver and gold bars somewhere in the wilds of Chihuahua. Taking a secret map of the location to a local tattoo artist, Villa ordered the man to shave his head and tattoo the map on his bald skull. Afterwards, he burned the map and shot the unfortunate tattooer. Obviously, denizens of the local cantinas said, treasure hunters had dug up the body, chopped off the head, shaved it, and were now on their way to becoming millionaires when they located the treasure.

It was a great story, but not taken seriously when someone recalled they read a similar tale in the works of the Greek historian Herodotus from 500 B.C. In Book V of his *Histories*, Herodotus relates how the Greek tyrant Histiaeus, wishing to send a secret message calling for a revolt, shaved the head of a slave, inscribed the message on his bald head, let the hair grow out, and sent the messenger to his friend and cohort Aristagoras.

There was general agreement, however, that it took a man of great boldness to tempt the gods with such a sacrilegious act. Suspicion soon turned to the *gringo* adventurers who, fearing neither God nor man, had roistered through Mexico during its bloody decade of revolution.

Such a man was Tracy Richardson, who had fought on many sides during the revolution, most bitterly against Villa during the Orozco rebellion of 1912. Richardson was known to be in Mexico and hated Villa. If the price was right, he was capable of anything. He was not above hawking Villa's head in any bazaar in the United States or Mexico. Richardson, however, was said to be conducting a forest survey in the state of Chiapas more than a thousand miles away.

Another name that immediately came to mind was that of Sam Dreben. The police recalled that when the Mexican revolution broke out, Dreben joined General Pascual Orozco and his hell-raising rebel "Colorados" before being defeated by the government forces of President Francisco Madero. More significantly, Dreben fought against Villa and scouted for General Pershing during the American punitive expedition in 1916.

After fighting with the U.S. Army in France during World War I, Dreben went into business in the border town of El Paso. He had recently intervened in a revolutionary outbreak in neighboring Juárez. There were rumors, however, that his business interests in El Paso were failing.

The Mexican police wondered if the intrepid Dreben attempted to recoup his fortune by stealing and selling the head of "The Centaur of the North?" Unfortunately, authorities learned that the previous March, tough Sam Dreben died in a Hollywood, California hospital, a victim of medical malpractice.

Another version of the story had it that when the police were speculating who had the courage to pull off such an atrocity, one police officer suggested, *"Es posible, el Colonel Holmdahl?"* The name brought a gasp from the assembled officials. Even in the years after the end of the revolution, Holmdahl was a name to conjure with. Police officers reported that he had been seen making the rounds of Parral bars the previous night.

Acting on a tip, the police rushed to the Hotel Casa Fuentes in Parral and arrested Holmdahl and Alberto Corral. Searching their automobile, they found a mysterious bottle, which, the police said, smelled like embalming fluid. They also confiscated a bloodstained ax, a large machete-like knife, and a shovel. There was never any mention of finding the gold bars in the trunk compartment; if indeed, they ever existed.

At the police station, Holmdahl was asked what he was doing in Parral and, more pointedly, where he had been the previous night. Smoothly, the American replied he was prospecting for copper deposits in the nearby mountains for his employer, the American Smelting and Refining Company. Maybe. That company, when queried by the author, stated that they held no employment records that far back in time. On the night in question, Holmdahl said he and his friends were relaxing by driving around and drinking at a number of Parral *cantinas*.

When a police officer held up the bottle, stated it was embalming fluid, and accused Holmdahl of using the stuff to preserve the severed head, Holmdahl was indignant. The bottle contained only mineral water, he explained. He told the officers that he had a serious kidney condition from drinking too much tequila. He said he constantly drank the water to ease his stomach pains. The police snorted at this and led Holmdahl and his two companions to cells in the Parral *calabozo* pending a hearing before a local judge. The bottle, a key bit of evidence, was put into custody at the police station.

Hours later, an American mining engineer, Bryan Brown, who knew Holmdahl, visited his cell and anxiously inquired if he could be of help. Holmdahl, rather smugly, told him, "Don't worry. I don't have the head and I'm fully protected."

The next morning, Holmdahl and Corral were hauled into court. When the police prosecutor testified that the bloody ax was probably used to chop off Villa's head, Holmdahl replied that since Villa had been dead for more than three years there would not be any blood. Besides, he said, when he and his friends were prospecting in the mountains they shot a deer, chopped it up, cooked it, and ate it. The same explanation held for the knife.

Confronted with the shovel, Holmdahl said it had been used to dig their car out of a ditch. He later told a newspaper reporter, "I had a hard time explaining how the fresh mud came to be on the shovel."[3] Then he was told their automobile was seen near the graveyard at 9:00 p.m. the night the grave was looted. "That had me stumped. We sure had driven near the graveyard but I explained our route from the mining deposits had taken me by the cemetery," he responded.

Throughout the interrogation, the presiding judge seemed disinterested in the proceedings until, with a flourish, the police official held up the bottle taken from Holmdahl's car. Placing the bottle on the evidence table, the prosecutor demanded to know why he was carrying embalming fluid in his car.

Holmdahl appealed to the judge, "Your honor it is only mineral water. I must drink it because I have a bad liver from drinking too much tequila." Striding to the evidence table, he picked up the bottle, saying, "If this is embalming fluid, drinking it will kill me." Before an astonished court, he lifted the bottle to his lips and in half a dozen large gulps drained it. "*Es verdad*, I am innocent," he announced. Perhaps.

The judge, impressed with this bravado performance, slammed down his gavel and pronounced, "Case dismissed. Release these men." Those who were cynical about the Mexican justice system, and there were many, later maintained it was not impossible that a police custodian, in receipt of a sizable gratuity, emptied the bottle of its poisonous contents and refilled it with water. And a judge who was *muy simpático* for unknown political reasons, might have eagerly found a reason to dismiss the case.

Holmdahl later told historian Bill McGaw that the judge said that he was worried about their safety, as the streets were full of outraged Villa supporters. When he offered a squad of local soldiers as an escort, Holmdahl, as an old Mexico hand, declined, saying, "Thanks but no thanks. The escort would just shoot us in the back." He added, "Just give us our guns back and we'll walk out of here." It was done and Holmdahl and Corral, with pistols conspicuously thrust in their belts, strode out of the courthouse.

With tight smiles, their hands hovering near their weapons, they stepped into the crowd and swaggered down the street to their hotel. No one made a threatening move, at least not then. The pair went to their hotel room, packed their bags, got into their auto, and headed out on the road to Juárez, where they would cross the border into El Paso and safety.

A few miles out of town, they stopped and opened the trunk compartment where they had stashed the gold bars. As expected, they had been stolen. Well, at least, they figured, they hadn't lost their lives. Villa supporters, however, according to Holmdahl, tried to ambush them on the road, but the old soldier snorted, "Though we were attacked several times between Parral and Juárez we got through. I'm still a pretty good shot."

Holmdahl later advanced the rather bizarre theory that the decapitation was planned by Plutarco Elías Calles, who had succeeded Alvaro Obregon as president of Mexico in 1924. A revolution was being planned by old Villa supporters, Holmdahl claimed, and the theft was designed to lure the plotters to Parral, where Calles would "disappear them." His theory convinced no one.

But if the culprits remained unknown, what could be the motive for purloining the head of the great man? Aside from the map on the head theory, many motives began to surface. One reputable Mexican historian, Elías L. Torres, in his book *La Cabeza de Villa*, reported that on the night of the desecration, an airplane landed on the small airstrip at Parral, located near the cemetery. Shadowy figures approached the plane, something was exchanged, and the aircraft took off, disappearing into the darkness. Torres claimed that an unnamed Mexican general who hated Villa's guts ordered the head cut off and delivered to him.

Today, Torres believes Villa's head, drilled out and used as a pen holder, sits on the general's desk. If so, it serves as a fitting rebuke for the semi-literate Pancho Villa. Another bolder Mexican historian, Oscar A. Martínez, named the man responsible as Brigadier General Francisco R. Durango, commanding officer of the garrison at Parral.

Soon a Mexico City newspaper, *El Gráfico*, printed a sensational story claiming the theft of the head was financed by an "eccentric Chicago millionaire," who planned to donate it to a scientific institution. Chicago and New York newspapers subsequently quoted Dr. Orlando F. Scott, a well-known Chicago brain specialist, who said he expected the head to arrive, "in a few days." Dr. Scott said the head would be examined from a pathological standpoint by experts from universities and hospitals. Dr. James Whitney Hall, an alienist and criminologist in Chicago, was quoted in the *New York Times* as expressing interest in studying the head.

The statements created an uproar among Mexican officials, who demanded the return of the head. The uproar prompted the American Medical Association to issue a press release stating the skull would be worthless from a scientific standpoint. In any event, the head never arrived in Chicago; or, if it did, no one admitted it.

Another rumor charged that the Ringling Brothers had purchased the head for $5,000 and planned to exhibit it in their circus freak show. The charge brought a scathing denial from John Ringling North. An old Yaqui Indian woman living in Los Angeles told a reporter that many of her tribe believed that Villa had made a pact with the devil. "If you will protect me in battle," Villa promised Lucifer, "I will give you my head after death." The devil finally got around to collecting his trophy, the woman said.

One story had it that the citizens of Columbus, New Mexico, had offered a reward of $50,000 for the head of Villa, dead or alive. It was discounted, however, when authorities realized that the entire population of the town couldn't raise even a tenth of that amount.

In later years, the *Los Angeles Times* quoted a Mrs. Gene Ernest as stating that in 1926 she was operating a store located in the Sheldon Hotel in El Paso. She recounted, "I became acquainted with Holmdahl at that time. I do not know whether he had Villa's head but he had something of great value, which he kept in his room. His Yaqui Indian guide slept in the room and guarded it at all times when Holmdahl was absent."[4]

Ben F. Williams, an El Paso cattleman and merchant, in his memoirs, *Let the Tail Go With the Hide*, published in 1984, wrote that he and Holmdahl were good friends who often dined together at El Paso's Central Cafe. Williams wrote that in March 1926, he and Holmdahl were having a number of drinks together when the mercenary said he had taken the head and was paid $25,000, plus expenses, for doing "the job."[5]

Years later, Williams was in Phoenix visiting a friend named Frank Brophy, a graduate of Yale University and a member of the university's Skull and Bones Club. Brophy told him that he and four other friends each put up $5,000 and hired Holmdahl to get the head. Holmdahl delivered it, and Pancho Villa's battered skull, according to Brophy, is now lodged in the trophy room at the club's headquarters at Yale.[6]

With no real clues, speculation finally died down, until a friend of Holmdahl, L.M. Shadbolt, revealed that in 1928 he met the soldier of fortune in El Paso. Shadbolt said Holmdahl entered his room in the Sheldon Hotel, unwrapped a bundle of newspapers and out rolled Villa's head. "I'm going to get $5,000 for it," Holmdahl said.

In 1932, however, a local El Paso historian, Larry A. Harris, said a reliable friend telephoned him saying Tracy Richardson, dodging Mexican police, had crossed the Rio Grande and brought the head to El Paso. Waiting to collect $10,000 for the trophy, Richardson, the man said, buried it in the nearby Franklin Mountains for safekeeping. At this news, a wave of curious diggers churned up Mount Franklin during the ensuing weeks, but to no avail.

Little was heard about the head until 1952, when the United States Secret Service located Emil Holmdahl living in retirement in Van Nuys, California. They questioned him about a horde of gold bars reputedly dug up in Mexico and illegally brought to the United States. The gold, the Secret Service said, was rumored to have been found buried in Mexico as a result of a map found on the skull of Pancho Villa. Nothing came of the investigation.

Nothing more was heard about the skull until it became an issue in the presidential election year of 1988 when Vice-President George Bush, running for president of the United States, was accused of knowing its whereabouts. Bush, a Yale alumnus, was a member of that university's Skull and Bones club which, it was said, had a collection of skulls of both the famous and infamous on display in their clubhouse.

Also, Bush was not the first member of his family to be involved with purloined skulls. His father, Prescott Bush, a former senator from Connecticut, reputedly was involved in digging up the body of the murderous Apache raider, Geronimo, cutting off his head, and ensconcing it in the Skull and Bones club. In between pronouncements on the economy and American foreign policy, Bush denied any knowledge of knowing the whereabouts of Villa's skull.

Historian Friedrich Katz adds a few more suspects to the list. In addition to Holmdahl's suspicious presence in Parral, he reports the rumor that a Colonel Durazo gave the order to some of his men to cut off the head and give it to Mexican President Obregón "who wanted Villa's skull for himself." Another general was said to want the head examined by scientists "to determine why he [Villa] was such a military genius." Katz concludes, however, there is really no hard evidence to substantiate any of these stories.[7]

Holmdahl, for the rest of his life, continued to deny he was guilty. But for all the denials, the most plausible scenario of the deed is that he cut off the head and delivered it to the pilot of the airplane landing at Parral the night of the decapitation. Carried to Mexico City, an ancient enemy of Villa probably spent his declining years looking at the skull with two bullet holes in it and chuckling. Where truth ends and fantasy begins is anybody's guess. As they say on the border, *Quién sabe*? Who knows?

Today, along the Rio Grande, some old ones believe that on dark nights a ghostly figure with a headless body, wearing the tattered uniform of a Revolutionary General, can be seen roaming up and down the banks of the river. It is Villa, they say, blindly searching for his stolen head.

In later years Holmdahl worked for American petroleum companies exploring for oil in Sonora, Mexico. — Holmdahl Papers

15
Keep Coming On

The strong men keep coming on,
They go down shot, hanged, sick, broken.
They live on fighting, singing, lucky as plungers.
Call hallelujah, call amen, call deep thanks.
The strong men keep coming on.
— Carl Sandburg

The accusation that he had been the culprit who chopped off Villa's head dogged Holmdahl for years. In Chihuahua, aging veterans who had ridden with Villa sat around flickering campfires softly singing the ballad of *La Decapitación*, which accused Holmdahl of stealing the head for money. Some vowed to kill him.

A friend of Holmdahl's told William C. Stewart, a Los Angeles reporter, "Don't use my name. I don't want to be gunned down if I ever go back to Mexico." He then related that he and Holmdahl and two other men were drinking in a Mexican *cantina*, when a woman slipped over to their table and warned them that old Villa comrades were outside with machine guns. They were, she said, going to shoot Holmdahl and his friends when they left the bar. "Holmdahl just shrugged and stayed there. But the rest of us were made of different stuff." The other three ran out the back door, called a taxi, and beat it across the border. What happened after that is not known; Holmdahl, however, emerged unscathed.

"Revolutions are scarce," Holmdahl complained to Stewart during an interview, "and those you do find have all the sportsmanship drained out of them by graft. It's getting to be like prize fights, no patriotism anymore, everyone wants to be president." Reminiscing years later, Holmdahl told Stewart,

> *Fifty years ago a man could go to Mexico or Central America and take his pick of a dozen wars, insurrections or marauding expeditions. But the rules changed and soldiers of fortune have to admit that free-lance fighting is a thing of the past. The world has gone to hell.*[1]

While things were still cooling off in Mexico, Holmdahl, at loose ends, gave a series of interviews to newspaper reporters. Being interviewed about his adventures continued throughout his life. Now in his forties, he expounded on his many adventures, waxed philosophic about Mexico, talked of joining foreign armies, sometimes promoted some project or scheme, and enjoyed the attention.

In many of his interviews, the facts were badly skewed either because he exaggerated, or, more likely, because of the ignorance of the reporters who garbled names and places in improbable sequences. In the summer of 1926, he announced he was going to join the forces of Abd-El-Krim, whose Riff guerrillas were fighting the French in Morocco. "Of course the Riffs will be defeated, but I've always liked fighting with the underdogs," he said.[2]

On June 29, 1926, the *El Paso Herald* quoted Holmdahl,

> *I may go to Damascus to fight with the Druses against the French. I had a mysterious note* (now in the Holmdahl papers) *it said, 'If you want to take up the case of the underdog . . . go to the Druse mountains in Syria . . . and see some real fighting between the French and the Druses.' I was figuring on going to fight with the Riffs but they've quit. I don't know where I'll go next . . . Life is all a gamble anyway. I may get killed some of these days fighting but at that I'd rather go that way than be run down by a flivver.*[3]

When the reporter asked if Holmdahl had been involved in the alleged kidnaping of famous California evangelist Aimee Semple

McPherson, he enigmatically replied, "Well, maybe I did and maybe I didn't." (He didn't. McPherson's kidnaping was a farce. She faked the dramatic story of her abduction and escape from Mexico to cover up a sexual tryst with a member of her flock.) When a reporter asked, "Why haven't you ever married"? Holmdahl replied, "I am to those." He pointed to his knife, pistol, and cartridges.[4]

On April, 7, 1927, while in Brownwood, Texas on business, Holmdahl told a reporter,

> *I planned on going to the defense of Shanghai against the Cantonese, but when the Americans and British stepped in I decided not to go. Before that I expected to ship for Nicaragua but halted when American Marines were sent to that war torn country. For I intended to take sides with the rebels and I couldn't fight my own countrymen.*

The reporter described Holmdahl as "Tall, grey haired and erect as a pine, despite his 44 years spent among battle and bloodshed."[5]

According to Gordon Holmdahl, sometime during 1929 the aging soldier married a woman named Ann. She apparently was a lady who desired a settled, well-to-do type of homelife and within a very short time the marriage failed.[6] On January 4, 1930, Holmdahl made the news again when a *Phoenix Gazette* headline reported, "ADVENTURER RUNS RUM TO GET THRILL — EMIL HOLMDAHL SAYS HE WAS TIRED OF SITTING ALONE IN HOTEL." The story reported that, while Prohibition was in effect, Holmdahl, restless and wandering, was arrested for transporting 254 pints of whiskey and fifty pints of beer. Following his detention a hundred miles east of El Paso, he denied being a bootlegger, claiming instead that boredom, not money, was the motive for his action and that the whole episode was a thrill-seeking stunt.[7]

Through the remainder of the 1920s, Holmdahl was involved in mining and real estate ventures in Mexico. In January 1930, he gave an interview in which he said, "Mexico is entering an era of constructive development and economic advancement." Philosophically he remarked,

> *Soldiers of fortune and kindred adventurers make plenty of money at various times and lose all at others. But the fascination of action has never lessened in my many vivid years of dangerous enterprise.*[8]

Perhaps by the 1930s, Holmdahl believed his time was running out, but it was not his life that was ending, but his era. One portent was a newspaper story about an old comrade in arms, Tracy Richardson, who had fallen on hard times.

In July 1932, Richardson was indicted on charges of using the mails to promote a fraudulent Mexican gold-mining scheme in 1930. The six-count, 29-page indictment told a story of lost Spanish mines and fabulous wealth. The lost mine was described as having been hidden near Mexico City for more than a century. It was recounted that rich deposits of gold and nuggets "as large as the end of your finger" lay beneath eleven Mexican waterfalls. Richardson was arrested on January 31, 1932, in Denver, having forfeited a $1,000 bond by not appearing in federal court the previous December.[9]

It was Richardson's third brush with the law since the end of World War I. One charge was for "rudely displaying and flourishing a deadly weapon." The other, in 1922, was for murder of a man in New Orleans. Some of the machine-gunner's luck still held, and he beat all three charges. He proved that he managed a real gold mine in Mexico to dispel the fraud charge. He produced a permit entitling him to legally carry a gun, and he was "no-billed" on the murder charge after proving self defense. Still, he was often broke, and at forty-one years of age, his swashbuckling image was beginning to fade.[10]

In May 1932, Holmdahl became involved in a new venture with Garret Peck, an inventor living in Hollywood, who claimed to be a graduate of MIT. Peck maintained he had developed a new principle of dirigible airship propulsion. The two embarked on a Hollywood to New York trip, presumably to raise money for the construction of the radical lighter-than-air craft. Holmdahl's job apparently was to garner publicity by giving interviews about the project and promoting a dirigible flight around the world. As

Holmdahl dazzled reporters with tales of high adventure far and wide; Peck talked technology.

In his projected airship, Peck boasted

> *Air is sucked into the craft by two propellers through a tube running through the center of the ship and is expelled by two pusher-type propellers. Steering, raising and lowering are controlled by a universal joint arrangement.*[11]

The idea of running the tube through the center of the aircraft was to decrease the resistance caused by the blunt end of the airship. Peck explained that the tube would do away with 87½ percent of the resistance encountered by the Akron. (*The Akron*, a U.S. naval airship completed in 1931, crashed on April 4, 1933, in a storm off the New Jersey coast, killing all 73 crewmen.)

The inventor said his new design would increase speed, driving the behemoth from 350 to 700 miles an hour and could carry a passenger from New York to Los Angeles for $30.00, or less than one cent a mile. During their joint interview, Holmdahl chimed in that after completing their flight around the world, he was going to settle down and spend his time constructing dirigibles. Peck said he had allowed Sir Hubert Wilkins, the noted Australian polar explorer, to use his patent on a submarine he was planning to construct. Patents, Peck said, had been obtained in every country except Russia.[12]

Luckily, for both of them, the Peck dirigible was never built. There were not enough naive investors in the Depression-wracked United States to invest in such a project. Had there been, they would, undoubtedly, have suffered the same fate as every other rigid airship from the *Zeppelins* of World War I to the *Hindenburg* of the Nazi era.

Now divorced from his first wife, Holmdahl married again. Elizabeth, called Betty by his family, was a pleasant, outgoing woman. She accepted Holmdahl's repeated forays into Mexico and assorted "get-rich-quick" schemes with equanimity, according to his nephew, Gordon Holmdahl. With a daughter, Ramona, from Betty's first marriage, the three settled in Van Nuys, California, their home life broken only by Holmdahl's business ventures into Mexico.[13]

In October 1934, Fortino Contreras, a well-known Mexican composer and old friend of Holmdahl's, wrote a military march entitled "Soldado de Fortuna." He dedicated it to Holmdahl in honor of his heroism during the Mexican revolution. The old soldier obtained a U.S. copyright for the composition on March 28, 1939.[14]

Throughout the remainder of the 1930s and into the 1950s Holmdahl did scouting work for petroleum companies, prospected for minerals, and engaged in real-estate promotions in Mexico. With the outbreak of World War II, he applied for active duty with the United States Army, but was turned down in a cursory manner.[15]

Because Holmdahl's service as a scout for Pershing was either overlooked or unknown to the Mexican government, or possibly because Villa was not held in high repute by government leaders in Mexico City, Holmdahl remained on excellent terms with many of the Mexican generals with whom he had served during the revolution. The *Historical and Biographical Dictionary of the Mexican Revolution* published in Mexico City, relates that ". . . in 1952 the Mexican Government made him a member of its Legion of Honor and at the same time gave him the honorary rank of Colonel for his service during the Revolution."[16]

The ceremony was held in Waterfill, Chihuahua, just across the Texas border a few miles east of El Paso, and it was quite a festive affair. A military band played the Contreras march, and many veterans of the revolution sang the martial *corridos* of the day, as their aging *compadre* was presented with his honorary commission. The beer and tequila flowed copiously, but it is doubtful that they sang the *Decapitación*.

Later, in 1952, Secret Service Agents knocked on the door of his Van Nuys home. It seems, the agents told Holmdahl, there was a rumor that someone had smuggled $20,000,000 in gold ingots across the Mexican border, and Holmdahl was the chief suspect.[17] Holmdahl denied any knowledge of such a crime, but one wonders, was the old soldier still involved with finding Villa's gold? The investigation, however, was subsequently dropped.

In 1957, his wife Elizabeth suddenly died. She was buried in Forest Lawn Cemetery, and Holmdahl moved in with his stepdaughter, Mrs. Ramona L. Foster.[18] In August 1958, Holmdahl was involved in his biggest project yet, the development of a gigantic resort community on Punta Banda peninsula. The land fronted Todos Santos Bay, approximately ninety miles south of San Diego on the west coast of Baja California, only a few miles from tourist developments at Ensenada.

The area offered duck hunting, deep-sea fishing, skin diving, and a warm, pleasant climate ideal for retirees. The development was to have modern docking facilities, charter-boat availability, a mobile-home facility as well as a hotel resort, and homes built to the owner's specifications. The project was to be developed by a consortium of Mexican and American businessmen, and would be, their advertisements said, "The Riviera of the West." Holmdahl's contribution to the project was apparently his inexhaustible number of governmental and business contacts in the area.[19]

Emil Holmdahl died at age seventy-nine while preparing for an oil exploration trip into Yaqui country.
— Holmdahl Papers

When historian Bill McGaw interviewed Holmdahl in June 1962, Holmdahl told McGaw he was heading back to Mexico, where he was organizing the settlement of a huge land grant in Baja, for settlement by French refugees fleeing from newly-independent Algeria.[20]

During the last decade of his life, Holmdahl must have missed his roistering comrades. In 1925, Sam "The Fighting Jew" Dreben, his health shattered and flat broke, died at the hands of a clumsy nurse who gave him the wrong injection. Tracy Richardson, "The World's Greatest Machine Gunner," served as a lieutenant colonel in the U.S. Army during World War II. But after the war he was reduced to selling household goods door-to-door, until he died broke in 1949.

Lee Christmas, his massive bulk reduced to ruin by an assortment of tropical diseases, died raving in a New Orleans hospital in 1924. Edward "Tex" O'Reilly, who fought under eight flags from the Philippines to China to the Equator, died in a veteran's hospital in 1946. Holmdahl was alone. He was the last of that breed of soldiers-of-fortune who fought their way from the Philippines to Mexico under an assortment of flags, mostly foreign.

His nephew, Gordon Holmdahl of Dublin, California, remembers the old soldier with great affection. Gordon Holmdahl is in possession of the original pardon papers signed by President Woodrow Wilson, as well as various swords, pistols, and other memorabilia of his uncle's fantastic career. "Uncle Emil," he said, "used to keep us spellbound for hours with stories of his adventures. I still have a photograph of him riding his horse with a little dog perched on his saddle. Emil said he carried him into battle with him."

Holmdahl, according to his nephew, "spoke the Yaqui language like a native." He supported himself in his later years as a prospector for American mining companies in Mexico, and even into his seventies made regular trips into remote areas of Mexico to bring ore samples back to the U.S.

According to Gordon Holmdahl,

> During the last year of his life, when he was in poor health, he was planning a trip back into some remote area of Mexico. My Dad and I tried to talk him out of it, we even tried to hide his car keys, but he found them. On April 8, 1963, while loading his automobile with his prospecting tools, he suffered a sudden massive stroke. He died almost instantly. He was buried in a crypt alongside his wife of almost thirty years. He was nearly 80 years old.[21]

If "Taps" were played at Holmdahl's funeral, it sounded not only for the old veteran, but for an age when soldiering could still be an adventure. He was the very last of the swashbuckling soldiers of fortune. It's difficult to come up with a fitting epitaph for a man like

Holmdahl. He served his country bravely and honorably in the Philippines as a foot-slogging infantryman, with Pershing and Patton as a daring scout, and as an officer fighting with the American Expeditionary Force in France during World War I. He was respected by honorable men on both sides of the border.

As for the rest, perhaps the words of the Canadian adventurer-poet Robert Service might well be appropriate.

Yes we go into the night as brave men go
we're hard as cats to kill,
And our hearts are reckless still,
And we've danced with death a dozen times or so....
Of our sins we've shoulders broad to bear the blame;
But we'll never stay in town and we'll never settle down ...
No, there's that in us that time can never tame;
And life will always seem a careless game . . .

Notes

Preface

1. Inscription in Holmdahl Family Bible and birth certificate in possession of Gordon Holmdahl, Dublin, California.
   Conversations with Gordon Holmdahl. April, 1997.
2. *The Saturday Blade*, Chicago. December 13, 1913. Interview with Emil Holmdahl.

Chapter 1

1. McGaw, William C. *Southwest Saga*. Phoenix: Golden West Publishers, 1988. 98.
2. Constantino, Renato. *Filipiniana*. Manila, Philippines: Cacho Hermanos, Inc, 1973 Reprint. 261.
3. Boller, Paul F., Jr. *Congressional Anecdotes*. Oxford and New York: Oxford University Press, 1991. 243.
4. Miller, Stuart Creighton. *Benevolent Assimilation*. Cambrige: Yale University Press, 1982. 112.
5. Ibid. 52-53.
   Blount, James H. *The American Occupation of the Philippines, 1898-1912*. New York: Oriole Editions, 1973. 155.
6. Ibid. 155.
7. Bain, David Howard. *Sitting in Darkness: Americans in the Philippines*. Boston: Houghton Mifflin Company, 1984. 81.
8. Sexton, William Thaddeus. *Soldiers in the Sun: An Adventure in Imperialism*. Freeport, New York: Books for Libraries Press, 1971. 111-112.
9. Bain. *Sitting In Darkness*. 82.
10. Faust, Karl Irving. *Campaigning in the Philippines*. New York: Arno Press and *The New York Times*, 1970. 151-152 and 178-183.
11. Ibid. Fighting around San Fernando. 187-191.
12. Miller. *Benevolent Assimilation*. 187.

13. Biographical Sketch of Emil L. Holmdahl received from the Consuládo General De México, El Paso, Texas, 1997.
14. Clipping from the *St. Paul Dispatch*. Holmdahl Papers. Bancroft Library of The University of California at Berkeley. July 23 c. 1900.
15. *The Saturday Blade*. Chicago. December 13, 1913.
16. Holmdahl Papers. 20th Infantry. Records of Service of Emil L. Holmdahl. Holmdahl qualified as a sharpshooter for the years 1904-05-06. His records show many hits on targets which simulated men sitting or lying prone. The hits were made at 1,000 yards.
17. Miller, *Benevolent Assimilation*. 179.
18. Sawicki, James A. *Infantry Regiments of the United States Army*. Wyvern Publications, 1981.
    Annual Report of the War Department 1901-1906. Washington, D.C.
19. Hahn, Emily. *The Islands*. New York: Coward, McCann & Geoghan, 1981.
20. Sexton, *Soldiers of the Sun*. 273.
21. Miller, *Benevolent Assimilation*. 189.
22. Sexton, *Soldiers in the Sun*. 240-241.
    Leckie, Robert. *The Wars of America*. New York: Harper-Collins Publishers, 1992. 570.
23. Constantino, *Filipiniana*. 101, 102.
24. Constantino, *Origin of a Myth*. 17.
25. Karnow, Stanley. *In Our Image: America's Empire in the Philippines*. New York: Random House, 1989. 174.
26. Leckie. *The Wars of America*. 574.

Chapter 2

1. Wocester, Dean C. *The Philippines Past and Present*. New York: The McMillan Company, 1930. 495.
2. Bacevich, Major Andrew J. Jr. "Disagreeable Work: Pacifying the Moros, 1903-1906." *Military Review*. Vol. LXII. No 6. June, 1982. 53.
3. Ibid. 53-54.
4. Scott, Major General Hugh Lenox. *Some Memories of a Soldier*. New York and London: The Century Company, 1928.
   314.
5. Ibid., 287.
6. Ibid., 313-314.
7. Ibid., 315-316.
   The author's father, Vincent D. Meed, was a lieutenant assigned to the 31st Infantry Regiment stationed in Luzon during the early 1920s. He recounted that he and several other officers wearing civilian clothes were shopping in the market town of Baguio, when a *juramentado* leaped from the jungle. Greased, his body bound with vines, foaming at the mouth and swinging a bolo, he charged one of the officers. The American officer whipped out a

.380 caliber revolver from under his coat and opened fire. The *juramentado* absorbed the bullets and practically severed the head of the officer before he fell dead. Examining the corpse of the dead fanatic, the American officers found five bullet holes in the man's chest. After that, my father said, no matter how uncomfortable it was, he put away his .38 and carried only the regulation Colt .45 caliber automatic pistol.

8. Holmdahl Papers. Notice of Promotion to Corporal. 20th Infantry Regiment. March 6, 1905.
9. Ibid. Memo certifying Holmdahl proficient in Infantry Drill Regulations. September 7, 1905.
10. Ibid. Undated, unnamed newspaper clip.
11. Scott. *Some Memories of a Soldier.* 328-331.
12. Foreman. *The Philippine Islands.* 584-585.
13. Welch, Richard E. Jr. "American Atrocities in the Philippines: The Indictment and the Response." *Pacific Historical Review.* No. 43. May, 1974. 233.
14. Bacevich. "Disagreeable Work." 59.
15. *Army News.* April 6, 1906.
16. Ibid.
17. Holmdahl Papers. Orders. Headquarters 20th Infantry Regiment. Zamboanga, Mindanao, Philippine Islands. December 15, 1905.
18. *Army News.* April 6, 1906.

Chapter 3

1. Institute of Oral History. University of Texas at El Paso. Interview with Brig. Gen. S.L.A. Marshall. No. 181. Interview by Richard Estrada. January 5,7,9,11,19, 1975. 37.
2. McGaw. *Southwest Saga.* 98.
Conversations with Gordon Holmdahl.
3. Snyder, Louis L. & Richard B. Morris, Eds. *A Treasury of Great Reporting.* New York: Simon & Schuster, 1949. 268.
4. Ibid.
5. *Army News.* April 26, 1906.
6. Ibid.
7. Ibid.
8. London, Jack. *Collier's Weekly Magazine.* May 5, 1906.
9. Ibid.
10. Holmdahl Papers. Handwritten Diary Page. April 18-26, 1906.
11. Ibid.
12. *Army News.* December 6, 13, 1906.

Chapter 4

1. Holmdahl Papers. Handwritten Diary. 2.
2. Davis, Richard Harding. *New Orleans Daily Picayune.* October 27, 1892. 3:6.
3. Langley, Lester D. & Thomas Schoonover. *The Banana Men.* Lexington: The University Press of Kentucky, 1995. 63.
4. Ibid. 39.
5. Ibid. 99.
6. Ibid. 99.
7. Ibid. 85.
8. Meed, Douglas V. *Bloody Border.* Tucson: Westernlore Press, 1992. 75.
9. Ibid. 86.
10. Davis, Richard Harding. *Notes of a War Correspondent.* New York: Charles Scribner's Sons, 1911. 189, 190.
11. Leibson, Art. *Sam Dreben—"The Fighting Jew".* Tuscon: Westernlore Press, 1996. 57.
12. *The Saturday Blade*, Chicago. December 13, 1913.
13. Langley & Schoonover, *Banana Men.* 68.
14. Leibson. *Sam Dreben.* vii-viii.
15. Richardson, Tracy. "A Soldier of Fortune's Story." *Liberty Magazine.* October 10, 1925.
16. For accounts of the adventures of Sam Dreben, Tracy Richardson, Tex O'Reilly and other soldiers of fortune of that era see:
    Meed, Douglas. *Bloody Border.* Tucson: Westernlore Press, 1992.
17. Richardson, Tracy. "A Soldier of Fortune's Story."
18. Langley & Schoonover. *Banana Men.* 190.
19. *El Paso Times.* December 8, 1913.

Chapter 5

1. Krauze, Enrique. *Mexico: Biography of Power: A History of Modern Mexico, 1810-1996.* New York: HarperCollins Publishers, 1997. 240.
2. Holmdahl Papers. "As a Soldier of Fortune and Filibuster in Mexico." (Hereinafter referred to as Filibuster.) 1.
3. Ibid.
4. For the Cananea strike see *Bloody Border*, 1-32.
   For background on Flores Magón and the radical movement, see: Hart, John M. *Anarchism and the Mexican Working Class, 1860-1931.* Austin: University of Texas Press, 1987.
5. Flores Magón, Ricardo and David Poole, eds. *Land and Liberty: Anarchist Influences in the Mexican Revolution.* Sanday: Cienfuegos Press, 1977. 46.
6. Ibid. 63.
7. Filibuster. 2.
8. Holmdahl Papers. Handwritten manuscript entitled "Bloody Mexico." undated.

9. Filibuster. 2.
10. McGaw Interview. Holmdahl Papers. There is a memo on the letterhead of Ferrocarril De Sud-Pacífico de México dated December 11, 1909 which states: "To Whom it May Concern: This is to certify that the bearer Mr. E.L. Holmdahl has worked for the Commissary Department since July 10, 1909 and is leaving the service of his own accord, furthermore beg to state that his services were very satisfactory. W.C. Dunn Jr."
11. Letter: E.L. Holmdahl to Adjutant General U.S. Army. December 24, 1913. National Archives.
12. Historical and Biographical Dictionary of the Mexican Revolution. International Section. Instituto Nacional de Estudios Históricos de La Revolución Mexicana, Mexico City.
13. Mason, Gregory. "The Mexican Man of the Hour." *Outlook Magazine*. June 6, 1914. 302.
14. Perez, Esther R. & James & Nina Kallas, eds. *Those Years of the Revolution: As Told by Veterans of the War*. San José, California: Aztlan Today, 1974. 134.
15. Ibid. 194.
16. McGaw Interview.
17. Filibuster. 10.
18. Mexican Consulate at El Paso Biographical Sketch of Emil Holmdahl.

Chapter 6

1. DuToit, Brian M. *Boer Settlers in the Southwest*. El Paso: Texas Western Press. Southwestern Studies No. 101. University of Texas At El Paso, 1995. 2-18.
2. Garibaldi, Giuseppe. *A Toast to Rebellion*. New York: The Bobs-Merrill Company, 1935. 303.
3. Ruiz, Ramon Eduardo. *Triumphs and Tragedy: A History of the Mexican People*. New York and London: W. W. Norton & Company, 1992. 294.
4. Turner, John Kenneth. *Barbarous Mexico*. Austin: University of Texas Press, 1984 edition. 29.
5. Thord-Gray, I. *Gringo Rebel*. Miami: University of Miami Press, 1960. 90-101.
6. *Barbarous Mexico*. 31.
7. Knight, Alan. *The Mexican Revolution*. Boston: Cambridge University Press, 1986. Volume I. 336.
8. *The Daily Mexican*. Mexico City. March 2, 1912. A Luncheon Program for the event is in the Holmdahl Papers.

Chapter 7

1. Letter: E.L. Holmdahl to Adjutant General United States Army. March 4, 1912. National Archives.

2. Letter: Adjutant General United States Army to E.L. Holmdahl. August 23, 1912. National Archives.
3. Gilly, Adolfo. *The Mexican Revolution*. Thetford, Norfolk, Great Britain: The Thetford Press Ltd, 1983. 65.
4. Camin, Héctor Aguilar and Lorenzo Meyer, . *In the Shadow of the Mexican Revolution: Contemporary Mexican History, 1910-1989*. Austin: University of Texas Press, 1993. 22.
5. Womack, John Jr. *Zapata and the Mexican Revolution*. New York: Vintage Books, 1970 edition. 127.
6. Gilly. *The Mexican Revolution*. 69.
7. Dunn, H.H. *The Crimson Jester: Zapata of Mexico*. New York: Robert M. McBride & Company, 1933. 18.
8. Ibid. 24.
9. Gilly. *The Mexican Revolution*. 71.
10. Ibid. 83.
11. Filibuster. 10.
12. Letter: E.L. Holmdahl to Adjutant General United States Army. April 21, 1912. National Archives.
13. Ibid.
14. Ibid.
15. Filibuster. 10.
16. Ibid.
17. Ibid.
18. Ibid.
19. Ibid.
20. Ibid.
21. Brunk, Samuel. *Emiliano Zapata: Revolution and Betrayal in Mexico*. Albuquerque: University of New Mexico Press, 1995. 46.
22. Ibid. 76.
23. Ibid. 76.
24. Conversations with Gordon Holmdahl.
25. Ibid.
26. Ibid.
27. Author's examination of the pistol. Dublin, California, April, 1997. William "Buffalo Bill" Cody carried an earlier model of this pistol called the "Schofield." While Cody was guiding the Grand Duke Alexis of Russia on a buffalo-hunting trip in the American West, the Duke noticed the weapon, fired it and was so impressed he placed a large order of pistols for use by the Russian military. With the addition of a spur below the trigger guard for placement of the second finger, the "Schofield" became known as the "Russian Model." Did the pistol belong to Zapata? There is only supposition. It is doubtful Holmdahl would have misspelled Cuernavaca; the dubiously literate Zapata might have. Also, a pistol of such historical significance would be worth a considerable sum of money to collectors.

Holmdahl, although sometimes strapped for money, never attempted to sell it but kept it until his death when it was inherited by his brother and later his nephew, Gordon Holmdahl.

Chapter 8

1. Meyer, Michael C. *Mexican Rebel: Pascual Orozco and the Mexican Revolution 1910-1915*. Lincoln: University of Nebraska Press, 1967. 24.
2. Perez. *Those Years of the Revolution*. 120.
3. Camin and Meyer. *In the Shadow of the Mexican Revolution*. 28.
4. Ruiz. *Triumphs and Tragedy*. 324.
5. Meyer. *Mexican Rebel*. 68.
6. Krauze. *Mexico: Biography of Power*. 309.
7. *Outlook* June 6, 1914. 394.
8. Meed. *Bloody Border*. 95-97.
    Richardson, Tracy. "A Soldier of Fortune's Story." In *Liberty* October 31, 1925.
9. Ibid.
10. Ibid.
11. Filibuster. 11.
12. *Historical and Biographical Dictionary of the Mexican Revolution*. International Section.
    Filibuster.
13. *Monterrey* [California] *Mexican-American*. June 12, 1912.
14. Richardson. "Soldier of Fortune."
15. *Monterrey* [California] *Mexican-American*. June 12, 1912.
16. *La Semana Illustrada* (Weekly Illustrated news magazine). June, 1912.
17. Letter: E.L. Holmdahl to Adjutant General United States Army. August 9, 1912. National Archives.
18. *Historical and Biographical Dictionary*. Mexico City.
19. Holmdahl Papers. Kosterlitzky to Holmdahl. October 24, 1912.
20. Holmdahl Papers. Holmdahl's handwritten report dated December 28, 1912. This report tells of "Red Flagger" operations in El Paso. Addressee unknown but probably sent to both General Trucy Aubert and the U.S. Bureau of Investigation.
21. Ibid.
22. Holmdahl Papers. Letter from Agent [name undecipherable], Department of Justice, Bureau of Investigation, Douglas, Arizona, November 4, 1913, to L.L. Hall, El Paso.
23. Holmdahl Papers. Note from H.J. Temple to Holmdahl. September 9, 1911.
24. Katz, Friedrich. *The Secret War in Mexico: Europe, the United States, and the Mexican Revolution*. Chicago: The University of Chicago Press, 1981. 92.
25. Ibid. 97, 98.
26. Ibid. 105.

27. Ruiz. *Triumphs and Tragedy.* 327.
28. Ibid.
29. Letter: E.L. Holmdahl to Adjutant General United States Army. December 24, 1913. National Archives.
30. *El Paso Morning News.* December 8, 1913.

Chapter 9

1. Dobie, J. Frank. *Apache Gold and Yaqui Silver.* New York: Little, Brown & Company, 1938. 323.
2. Krauze. *Mexico: Biography of Power.* 306, 307.
3. Torres, Elías L. *Twenty Episodes in the Life of Pancho Villa.* Austin: The Encino Press, 1973. x.
4. Krauze. *Mexico: Biography of Power.* 308.
5. Torres. *Twenty Episodes in the Life of Pancho Villa.* x.
6. Camin and Meyer. *In the Shadow of the Mexican Revolution.* 42, 43.
7. Guzmán, Martín Luis. *Memoirs of Pancho Villa.* Austin: University of Texas Press, 1973. 100.
8. *El Paso Morning News.* December 8, 1913.
9. Guzmán. *Memoirs of Pancho Villa.* 100.
10. "Victorias Del General Villa."(Victories of General Villa) T.F. Serrano and C. Del Vando. El Paso, Texas, 1913. 63. This is a pamphlet written in Spanish in possession of the author.
11. Thord-Gray. *Gringo Rebel.* 36-38.
12. *San Francisco Call.* November 25, 1913.
13. Holmdahl Papers. Unidentified newspaper clipping.
14. *San Francisco Call.* November 25, 1913.
15. Katz, Friedrich. *The Life and Times of Pancho Villa.* Stanford: Stanford University Press, 1998. 228.
16. Ibid.
17. *The Creston Advertiser-Gazette.* April 4, 1914.
18. *The Saturday Blade.* Chicago. December 13, 1913.
19. Richardson. "A Soldier of Fortune's Story."
20. Institute of Oral History. Number 181. University of Texas at El Paso. Interview with Brig. Gen. S.L.A. Marshall. Interviewed by Richard Estrada. July 5,7,9,11,19, 1975.
21. Ibid.
22. Letter: E.L. Holmdahl to Adjutant General United States Army. December 24, 1913. National Archives.
23. Ibid.
24. Torres. *Twenty Episodes in the Life of Pancho Villa.* 8-12.
25. Ibid.
26. *New York Times.* June 20, 1914. 2:3.
27. Katz. *The Life and Times of Pancho Villa.* 331.
28. Ibid.

29. Ibid. 343.
30. Ibid. 388.

Chapter 10

1. Reed, John. *Insurgent Mexico.* Middlesex, England: Penguin Books, Ltd., 1983. 69
2. *El Paso Times.* October 14,15,17,19,22, 1915.
   El Paso Herald. October 14,15,18,19,20, 1915.
   New York Times. October 21, 1915.
   The background and discussion of the plot and trial are from the listed newspaper accounts.
3. Ibid.
4. Holmdahl Papers. Telegram. October 15, 1914. O.R. Seagraves to Holmdahl.
5. *El Paso Times* and *El Paso Herald* October 14-22, 1915.
6. Holmdahl Papers. Telegram. December 12, 1914. 10:50 a.m. Pearce Forwarding Co. To Holmdahl.
7. Ibid. Telegram. December 12, 1914. 1:58 p.m. Pearce Forwarding Co. to Holmdahl.
8. Ibid. Telegram. December 12, 1914. 2:23 p.m. M. Brennan to Holmdahl.
9. Ibid. Telegram. January 10, 1915. M. Brennan to Holmdahl.
10. Ibid. Letter. February 23, 1915. Brig. Gen. J.H. Hernandez to Mayor Tom Lea.
11. Ibid. Transcripts of six statements taken from huertista officers. December, 1915.
12. Camin and Meyer. *In the Shadow of the Mexican Revolution.* 50.
13. Martínez, Oscar J. *Fragments of the Mexican Revolution: Personal Accounts From the Border.* Albuquerque: University of New Mexico Press, 1983. 248, 249.
14. Ibid., 249.
15. *El Paso Times* and *El Paso Herald.* October 21-22, 1915.
16. Application for Commission in the United States Army. Form No. 88-4 (3 pages). E.L. Holmdahl to Adjutant General United States Army. December 29, 1915. National Archives.
17. Letter: Adjutant General United States Army to E.L. Holmdahl. March 28, 1916.
18. Clendenen, Clarence C. *The United States and Pancho Villa: A Study in Unconventional Diplomacy.* Port Washington, N.Y.: Kennikat Press, 1961. 70, 80, 84.
19. A full discussion of the Plan de San Diego can be found in the following books and articles:
    Gerlach, Allen. "Conditions Along the Border-1915: The Plan de San Diego." *New Mexico Historical Review.* July, 1968. 195-207.
    Hager, William M. "The Plan of San Diego: Unrest on the Texas Border in 1915." *Arizona and the West.* Winter, 1963. 327-336.

Harris, Charles H. III and Sadler, Louis R. *The Border and the Revolution.* Las Cruces: New Mexico State University, 1988. 72-98.
Katz, Friedrich. *The Secret War in Mexico.* 340-342.
Meed. *Bloody Border.* 115-134.
Rocha, Rodolfo. *The Influence of the Mexican Revolution on the Mexico-Texas Border 1910-1916.* Lubbock: Texas Tech University, 1961. [PhD Dissertation.] 319-325.
Sandos, James A. "The Plan de San Diego: War and Diplomacy on the Texas Border 1915-1916." *Arizona and the West.* Spring, 1972. 5-24.
20. Clendennen. *The United States and Pancho Villa.* 72.
21. Ibid., 211.
22. Ibid., 210.
23. Katz. *The Life and Times of Pancho Villa.* 550, 551.
24. Ibid., 551-555.
    Krauze. *Mexico: Biography of Power.* 328,329.
    Ruiz. *Triumph and Tragedy.* 334.
25. Tompkins, Colonel Frank. *Chasing Villa.* Harrisburg: Military Service Publishing Company, 1934. 58.
26. Ibid., 55, 56, 88.
27. Vandiver. *Black Jack.* 604.
28. O'Connor, Richard. *Black Jack Pershing.* Garden City, N.Y.: Doubleday & Company Inc., 1961. 87.
29. Toulmin, Colonel H.A. *With Pershing in Mexico.* Harrisburg: Military Service Publishing Company, 1935. 20-21.

Chapter 11

1. Vandiver. *Black Jack.* 605.
2. The story of the first Aero Squadron is told in the following:
   Clendenen, Clarence C. *Blood on the Border.* London: The Macmillan Co., 1969. 315-322.
   Mason, Herbert Malloy, Jr. *The Great Pursuit.* New York: Random House, 1970. 103-119, 121.
   Vandiver. *Black Jack.* 605, 660.
3. Harris, Charles H. III & Sadler, Louis R. *The Border and the Revolution.* Las Cruces: New Mexico State University, 1988. 17.
4. *The Southwesterner.* Vol. 2, No. 4. October, 1962. 13.
5. Leibson. *Sam Dreben.* 108-109.
6. Braddy, Haldeen. "Myths of Pershing's Mexican Campaign." *Southern Folklore Quarterly.* Volume XXVII. September, 1963. Number 3. 184.
7. Mason. *The Great Pursuit.* 132.
8. Clendenen, Clarence. *Blood On The Border: The United and States Army the Mexican Irregulars.* London: The Macmillan Company, 1969. 260,261.
9. Ibid., 262, 263.

10. Blumenson, Martin, Ed. *The Patton Papers*. Boston: Houghton Mifflin Company, 1972. 362.
    Meed, Douglas V. "Lieutenant Patton's Raid." *True West Magazine*. May 1997. 44-48.
11. Blumenson. *The Patton Papers*. 362.
12. Ibid., 363.
13. Ibid., 364-365.
14. Brittain, Don L. "A Civilian With Pershing In Mexico." *Password Quarterly*. Volume XIV. Number 2. Summer 1969. 50.
15. Blumenson. *The Patton Papers*. 367.
16. *New York Times*. May 23, 1916. 5.
17. Brittain. "A Civilian With Pershing in Mexico." 51.
18. Holmdahl Papers. Letter. May 20, 1916. George Patton to Whom It May Concern.
19. Ibid. Note. July 10, 1916. W.W. Reed, Captain, 6th Cavalry to Commander of the Guard.
20. Ibid. Letter. July 24, 1916. M.C. Shallenberger to Quartermaster at the Base, Columbus, N.M.
21. Ibid. Telegram. December, 1916. Major General Frederick Funston to Holmdahl.
22. Holmdahl Papers. Telegram. December 26, 1916. General Flynn. Hdqts 606 Brigade To Scout Holmdahl.
23. Meed, Douglas. "Suicide Charge at Carrizal" *True West Magazine*. Volume 40. Number 9. September, 1993. 20-25.
24. Vandiver. *Black Jack*. 667.
25. Ibid., 685-686

Chapter 12

1. Holmdahl Papers. Letter. Holmdahl to U.S. Attorney General. February 1, 1917.
2. Ibid. Letter. U.S. Marshal, Austin, Texas to Holmdahl. February 9, 1917.
3. Ibid. Letter. Col. J.A. Ryan to Holmdahl. February 17, 1917.
4. Ibid. Letter. Holmdahl to Col. J.A. Ryan. February 22, 1917.
5. Ibid. Letter. Rep. Jeff McLemore to Holmdahl. March 19, 1917.
6. Ibid. Letter. Maj. Gen. John J. Pershing, U.S. Army Commanding to Hon. Jeff McLemore. March 11, 1917.
7. Ibid. Letter. Maj. Gen. Pershing to Attorney General of the Army. March 14, 1917.
8. Ibid. Newspaper clipping undated.
9. Ibid. Letters. Asst. U.S. Attorney to Holmdahl. March 21, 1917.
   Asst. U.S. Attorney to Holmdahl. April 14, 1917.
10. Ibid. Letter. Mayor Tom Lea to U.S. Attorney General. April 21, 1917.
11. Ibid. Letter. Holmdahl to Frank Polk. Dept. of State. April 30, 1917.

Notes

12. Braddy, Haldeen. "Myths of Pershing's Mexican Campaign" *Southern Folklore Quarterly*. Volume XXVII. September, 1963. Number 3. 191.
13. *New York Times Magazine*. "The Machine Gun Man of the Princess Pats." October 31, 1915.
14. Meed. *Bloody Border*. 109-110.
15. Ibid. Telegram. Major Sam Robertson, Sixth Reserve Engineers to Rep. Jeff McLemore. June 6, 1917.
16. Ibid. Telegram. Major Sam Robertson to Frank Polk, Dept. of State. June 7, 1917.
17. Ibid. Letter. Holmdahl to F.C. Proctor. July 21, 1917.
18. Ibid. Letter. 1st Lt. H.L. Taylor to Adj. Gen. U.S. Army. July 17, 1917.
19. Ibid. Report of Physical Examination by U.S. Army. December 10, 1918.
20. *Oakland Tribune*. August 20, 1917.
21. Holmdahl Papers. Memo. Adj. Gen. G.W. Read to War Dept A.G.O. July 17, 1917.
22. Ibid. Letter. Holmdahl to Sen. Morris Sheppard. July 21, 1917.
23. Ibid. Letter. Holmdahl to Rep. Jeff McLemore. July 21, 1917.
24. Ibid. Letter. James A. Finch, Dept. of Justice to Holmdahl. October 17, 1917.
25. Ibid. Letter. Holmdahl to Joseph P. Tumulty, Sec. to President Woodrow Wilson. December 19, 1918.
26. Buchanan, John. *A History of the Great War*. Volume IV. New York: Houghton Mifflin Company, 1922. 86.
27. *Historical Report of the Chief Engineer Including All Operations of the Engineer Department American Expeditionary Forces 1917-1919*. Washington D.C.: Government Printing Office, 1919. 158.
28. Ibid., 161-162.
29. The story of Carey's Chickens is told in:
    Halsey, Francis Whiting. *The Literary Digest History of the World War*. Vol. 5. New York and London: Funk & Wagnalls Company, 1919. 9-15.
    Horne, Charles F., Ed. *The Great Events of the Great War*. Vol. VI. National Alumni Press. 1920. 68-79.
30. Hart, B.H. Liddell. *Reputations Ten Years After*. Boston: Little, Brown & Company, 1928. 141-142.
31. Historical Report of the Chief Engineer. 162.
32. Horne, Charles F., Ed. *Great Events of the Great War*. 77.
33. Historical Report of the Chief Engineer. 180-181.
34. Holmdahl Papers. War Department Telegram. Department of Engineers to Holmdahl. September 2, 1918.
35. Ibid. Letter. Holmdahl To Commanding Officer 97th Engineers, Camp Leach. November 23, 1918.
36. *Los Angeles Times*. February 26, 1967.
37. Conversations with Gordon Holmdahl. Dublin, California. April, 1997.

Chapter 13

1. Holmdahl Papers. Letter. A.J. Bruff, Asst. Director of Sales, U.S. War Department To Anglo-South American Bank, Ltd., Madrid. September 3, 1919.
2. Ibid. Letter. E.C. Morse, Director of Sales, U.S. War Department to Holmdahl. May 25, 1920.
3. Ibid. Newspaper clip from undated El Paso newspaper. c. early 1920's.
4. Wellman, Paul I. *A Dynasty of Western Outlaws*. New York: Bonanza Books, 1961. 276-280.
   Patterson, Richard. *The Train Robbery Era*. Boulder, Colorado: Pruett Publishing Company, 1991. 117, 118.
5. Stewart, William C. "Soldier's Ghost Walks With Villa's Head" *Los Angeles Times*. February 19, 1967.
6. Holmdahl Papers. Clipping from *San Francisco Examiner*. Undated, c. early 1920's.
7. Ibid. Los Angeles newspaper. Undated, c. 1920's.
8. Ibid. Land deed signed by Holmdahl, Jennings, and Hughes on February 3, 1926 in the town of Carmargo, state of Chihuahua.
9. Ibid. Unnamed newspaper clipping. Headline "Ex Bank Robber To Turn Miner." c. 1926.

Chapter 14

1. Southwesterner. Volume 2, Number 3. September, 1962.
2. Stories of Villa's missing head are told in detail in the following publications:
   New Yorker. "The Cabeza de Villa." November 27, 1989. 108-120.
   Braddy, Haldeen. "The Head of Pancho Villa." *Western Folklore*. Berkeley and Los Angeles: University of California Press. Volume XIX. January, 1960. Number 1. 25-33.
   Meed, Douglas V. "Who Stole Pancho Villa's Head?" *True West Magazine*. August, 1996. 14-19.
   Southern Folklore Quarterly. Volume 27. September, 1963. Number 3. 192.
   El Paso Times. February 15,19, 1989.
3. Many of the details of his interrogation, according to Holmdahl, were related in an interview in the *El Paso Times* February 15, 1926.
4. *Los Angeles Times*. February 15, 1967. 6.
5. Williams, Ben. *Let the Tail Go with the Hide*. El Paso: Mangan Books, 1984. 78-81.
6. Ibid. 266.
7. Katz. *The Life and Times of Pancho Villa*. 789.

Chapter 15

1. *Los Angeles Times*. February 19, 1967.
2. Holmdahl Papers. Unnamed, undated California newspaper clipping.
3. *El Paso Herald*. June 29, 1926.
4. Ibid.
5. *Brownwood* [Texas] *Bulletin*. April 7, 1927.
6. Conversations with Gordon Holmdahl.
7. *Phoenix Evening Gazette*. January 4, 1930.
8. Holmdahl Papers. Unnamed Phoenix newspaper. January 14, 1930.
9. *Fort Worth Star-Telegram*. January 31, 1932.
10. Walker, Dale L. *Mavericks - Ten Uncorralled Westerners*. Phoenix: Golden West Publishers, 1989. 80-81.
11. *Fort Worth Star-Telegram*. May 24, 1932.
12. Ibid.
13. Conversations with Gordon Holmdahl.
14. Register of Copyrights. Copyright Office of the United States of America, Library of Congress, Washington D.C. May 8, 1939. The original sheet music of "Soldado de Fortune" is in the possession of Gordon Holmdahl.
15. Holmdahl Papers. Letter. From Holmdahl to Army Adjutant General.
16. *Diccionario Historico y Biografico de la Revolucion Mexicana*. Seccion Internacional. Mexico City.
17. *Southwesterner*. Volume 2. Number 3. September, 1962.
18. *Los Angeles Times*. February 26, 1967.
19. Holmdahl Papers. Overall Plan For Development of Punta Banda. August, 1958.
20. Southwesterner. Volume 2. Number 3. September, 1962.
21. Conversations with Gordon Holmdahl.

✺ Bibliography ✺

Primary Sources

ARCHIVES

NATIONAL ARCHIVES. War College Division #4439-37. Papers of Emil Holmdahl.
Letters:
    Holmdahl to Adjutant General, U.S. Army. Dec. 24, 1913.
    Holmdahl to Adjutant General, U.S. Army. March 4, 1912.
    Adjutant General, U.S. Army to Holmdahl. Aug. 23, 1912.
    Holmdahl to Adjutant General, U.S. Army. April 21, 1912.
    Holmdahl to Adjutant General, U.S. Army. Aug. 9, 1912.
    Adjutant General to Holmdahl. (undated).
    Holmdahl Application for Commission in U.S. Army. Dec. 29, 1915.
    Adjutant General to Holmdahl. March 28, 1916.

BANCROFT LIBRARY, University of California at Berkeley. Holmdahl Papers.

MANUSCRIPTS, LETTERS, TELEGRAMS, NOTES

    Holmdahl, Emil. Handwritten Diary. April 18-26, 1906.
    Unpublished Manuscript. "As A Soldier of Fortune in Mexico."
    Memo. "Ferrocarril De Sud-Pacífico." July 10, 1909.
    Letter. Colonel Emilio Kosterlitzky to Holmdahl. Oct. 24, 1912.
    Report on Red Flagger Activity in El Paso. Dec. 28, 1912.
    Letter. Department of Justice, Bureau of Investigation to L.L. Hall, El Paso. Nov. 4, 1913.
    Note. H.J. Temple to Holmdahl. Sept. 9, 1911.
    Telegrams. Pearce Forwarding Company to Holmdahl. Oct. 15, 1914. 10:50 a.m. & 1:58 p.m.
    Telegram. O.R. Seagraves to Holmdahl. Oct. 15, 1914.

Bibliography

Telegram. M. Brennan to Holmdahl. Dec. 12, 1914 & Jan. 10, 1915.
Letter. Brig. Gen. J.H. Hernandez to Tom Lea. Feb. 23, 1915.
Transcripts of statements by six Huerrista officers. Dec., 1915.
Letter. Lt. George Patton To Whom It May Concern. May 20, 1916.
Note. W.W. Reed, 6th Cavalry To Commander of the Guard. July 10, 1916.
Letter. M.C. Shallenberger To Quartermaster at the Base, Columbus, N.M. July 24, 1916.
Telegram. General Flynn, Headquarters 606 Brigade to Scout Holmdahl. Dec. 26, 1916.
Letter. U.S. Marshal, Austin To Holmdahl. Feb. 9, 1917.
Letter. Col. J.A. Ryan To Holmdahl. Feb. 17, 1917.
Letter. U.S. Representative Jeff McLemore To Holmdahl. March 19, 1917.
Letter. Maj. Gen. John J. Pershing To Hon. Jeff McLemore. March 11, 1917.
Letter. Maj. Gen. John J. Pershing To Attorney General of the Army. March, 14, 1917.
Letter. Asst. U.S. Attorney To Holmdahl. March 21, 1917.
Letter. Asst. U.S. Attorney To Holmdahl. April 14, 1917.
Letter. Mayor Tom Lea To U.S. Attorney General. April 21, 1917.
Letter. Holmdahl To Frank Polk, Dept. Of State. April 30, 1917.
Handwritten Note. Undated. "Never Give Up."
Telegram. Major Sam Robertson, Sixth Reserve Engineers to U.S. Representative Jeff McLemore. June 6, 1917.
Telegram. Major Robertson to Frank Polk, Dept. Of State. June 7, 1917.
Letter. Holmdahl To F.C. Proctor. July 21, 1917.
Letter. Holmdahl To U.S. Rep. Jeff McLemore. July 21, 1917.
Letter. Holmdahl To U.S. Senator Morris Sheppard. July 21, 1917.
Letter. James A. Finch, Dept. of Justice To Holmdahl. Oct. 17, 1917.
Letter. Holmdahl to Joseph P. Tumulty, Secretary To President Woodrow Wilson. Dec. 19, 1918.
Unpublished manuscript. *Bloody Mexico* (undated).
Overall Plan for Development of Punta Banda, Baja Mexico. Aug., 1958.
Land Deed. Signed by Holmdahl, Jennings, and Hughes. Feb. 3, 1926 in town of Camargo, State of Chihuahua.

UNITED STATES ARMY RECORDS IN HOLMDAHL PAPERS

20th Infantry Regiment. Qualification as Sharpshooter. 1904-5-6.
Notice of Promotion To Corporal, 20th Infantry Regiment. March 6, 1905.
Certification, Cpt. Holmdahl proficient in 1904 Infantry Drill Regulations. Signed by Major Hugh Scott. Jolo, Philippine Islands. Sept. 7, 1905.
Orders. Headquarters 20th Infantry. Zamboanga, Mindanao, Philippine Islands. Dec. 15, 1905.
Report of Physical Exam by U.S. Army. Dec. 12, 1918.
Memo. Adj. Gen. G.W. Reed to War Dept. A.G.O. July 17, 1917.

Telegram. War Department, Dept. of Engineers to Holmdahl. Sept. 2, 1918.
Letter. Holmdahl to Commanding Officer 97th Engineers, Camp Leach. Nov. 23, 1918.
Letter. 1st. Lt. H.L. Taylor, U.S. Army Medical Corps to Adj. Gen. U.S. Army. July 17, 1917.
Letter. A.J. Bruff, Asst. Dir. of Sales, U.S. War Department to Anglo-South Bank, Ltd. Madrid. Sept. 3, 1919.
Letter. E.C. Morse, Dir. of Sales, U.S. War Department to Holmdahl. May 25, 1920.

GOVERNMENT DOCUMENTS
United States of America

Annual Report of the War Department 1901-1906. Washington, D.C. Government Printing Office.
Historical Report of the Chief Engineer Including All Operations of the Engineer Department American Expeditionary Forces 1917-1919. Government Printing Office. Washington D.C. 1919.
Register of Copyrights. Copyright Office of the United States of America. Library of Congress. Washington D.C. May 8, 1939. Original Sheet Music of "Soldado de Fortuna."

United States of Mexico

Consulado General de México, El Paso, Texas. Biography of Emil L. Holmdahl.
Historical and Biographical Dictionary of the Mexican Revolution International Section. Instituto Nacional de Estudios Historicos de La Revolución Mexicana. Mexico City.

NEWSPAPERS

Newspaper clippings without the name of the newspaper or the complete date are listed in the Holmdahl Papers. Identified newspapers are listed under Newspapers.

AMERICAN
Army News, Monterrey [California]
Bulletin, Brownwood [Texas]
Creston Advertiser-Gazette
El Paso Herald
El Paso Morning News
El Paso Times

Fort Worth Star-Telegram
Los Angeles Times
New Orleans Daily Picayune
New York Times
Oakland Tribune
Phoenix Evening Gazette
St. Paul Post Dispatch
San Francisco Call
San Francisco Examiner
Saturday Blade [Chicago]
The Southwesterner

MEXICAN
Daily Mexican [Mexico City]
La Semana Ilustrada
Monterrey Mexican-American [Monterrey, Mexico]

PHOTOGRAPHS AND MAPS

BANCROFT LIBRARY, University of California at Berkeley. Holmdahl Papers.
DOUGLAS V. MEED COLLECTION, In the possession of the author
EL PASO PUBLIC LIBRARY, El Paso Texas, Aultman Collection
GORDON HOLMDAHL COLLECTION, Dublin, California
HALCYON PRESS, Houston, Texas
NATIONAL ARCHIVES, Washington, D.C.
RADFORD SCHOOL, El Paso, Texas
SAN ANTONIO LIGHT Collection, San Antonio, Texas
SIERRA BLANCA MUSEUM, Sierra Blanca, Texas

DISSERTATIONS

Rocha, Rodolfo. *The Influence of the Mexican Revolution on the Mexico-Texas Border, 1910-1916.* Lubbock, Texas: Texas Tech University, 1981. Ph.D. dissertation.

INTERVIEWS, ORAL HISTORY AND FAMILY DOCUMENTS

Author's conversations with Gordon Holmdahl, Dublin, California. April, 1997.
Holmdahl Family Bible — inscription in possession of Gordon Holmdahl.
Author's conversations with Vincent D. Meed. Circa 1940s.
Institute of Oral History, University of Texas, El Paso. Interview with Brigadier General S.L.A. Marshall. No. 181. Interviewed by Richard Estrada. January 5,7,9,11,19, 1975.
Author's examination of the "Zapata" pistol. April, 1997.

Secondary Sources

BOOKS

Bain, David Howard. *Sitting In Darkness: Americans in the Philippines.* Boston: Houghton Mifflin Company, 1984.
Blumenson, Martin. Ed. *The Patton Papers.* Boston: Houghton Mifflin Company, 1972.
Blount, James H. *The American Occupation of the Philippines: 1898-1912.* New York: Oriole Editions, 1973.
Boller, Paul F., Jr. *Congressional Anecdotes.* New York and Oxford: Oxford University Press, 1991.
Brunk, Samuel. *Emiliano Zapata: Revolution and Betrayal in Mexico.* Albuquerque: University of New Mexico Press, 1995.
Buchanan, John. *A History of the Great War, Vol. IV.* New York: Houghton Mifflin Company, 1922.
Bunuan, Vicente G. *Arguments For Immediate Philippine Independence.* Washington, D.C: Philippine Press Bureau, 1924.
Camin, Héctor Aguilar and Lorenzo Meyer. *In the Shadow of the Mexican Revolution: Contemporary Mexican History, 1910-1989.* Austin: University of Texas Press, 1993.
Clendenen, Clarence C. *Blood on the Border: The United States Army and the Mexican Irregulars.* London: The Macmillan Company, 1969.
_____. *The United States and Pancho Villa: A Study in Unconventional Diplomacy.* Port Washington, NY: Kennikat Press, 1961.
Constantino, Renato. *Filipiniana.* Manila, Philippines: Cacho Hermanos, Inc., 1973 reprint.
_____. *Origin of a Myth.* Quezon City, Philippines: Malaya Books, Inc., 1968.
Davis, Richard Harding. *Notes of a War Correspondent.* New York: Charles Scribner's Sons, 1911.
Dobie, J. Frank. *Apache Gold and Yaqui Silver.* New York: Little, Brown and Co., 1938.
Dunn, H.H. *The Crimson Jester: Zapata of Mexico.* New York: Robert M. McBride and Company, 1933.
DuToit, Brian M. *Boer Settlers in the Southwest.* El Paso: Texas Western Press. Southwestern Studies No.101. University of Texas at El Paso, 1995.
Faust, Karl Irving. *Campaigning in the Philippines.* New York: Arno Press and *The New York Times*, 1970.
Flores Magón, Ricardo and David Poole, eds. *Land and Liberty: Anarchist Influences in the Mexican Revolution.* Cienfuegos Press. Sanday, 1977.
Foreman, John. *The Philippine Islands, 3rd Edition.* New York: Charles Scribner's Sons, 1906.
Garibaldi, Giuseppe. *A Toast to Rebellion.* New York: The Bobbs-Merrill Company, 1935.

Gilly, Adolfo. *The Mexican Revolution*. Thetford, Norfolk, United Kingdom: The Thetford Press Ltd, 1983.

Guerrero, Amado. *Philippine Society and Revolution*. Hong Kong: Ta Kung Pao, 1971.

Guzmán, Martín Luis. *Memoirs of Pancho Villa*. Austin: University of Texas Press, 1973.

Hahn, Emily. *The Islands*. New York: Coward, McCann & Geohegan, 1981.

Halsey, Francis Whiting. *The Literary Digest History of the World War, Vol. 5*. New York and London: Funk & Wagnalls Company, 1919.

Harris, Charles H. III and Louis R. Sadler. *The Border and the Revolution*. Las Cruces, New Mexico: New Mexico State University Press, 1988.

Hart, B.H. Liddell. *Reputations Ten Years After*. Boston: Little Brown & Co., 1928.

Hart, John M. *Anarchism and the Mexican Working Class, 1860-1931*. Austin: University of Texas Press, 1987.

Horne, Charles F., Ed. *The Great Events of the Great War, Vol. VI*. National Alumni Press, 1920.

Karnow, Stanley. *In Our Image: America's Empire for the Philippines*. New York: Random House, 1989.

Katz, Friedrich. *The Life and Times of Pancho Villa*. Stanford University Press, 1998.

_____. *The Secret War in Mexico: Europe, the United States, and the Mexican Revolution*. Chicago: The University of Chicago Press, 1981.

Knight, Alan. *The Mexican Revolution*. Cambridge: Cambridge University Press, 1986.

Krauze, Enrique. *Mexico: Biography of Power, A History of Modern Mexico, 1810-1996*. New York: Harper Collins Publishers, 1997.

Langley, Lester D. and Thomas Schoonover. *The Banana Men*. Lexington: University Press of Kentucky, 1995.

Leckie, Robert. *The Wars of America*. New York: Harper Collins Publishers, 1992.

Leibson, Art. *Sam Dreben—The Fighting Jew*. Tucson, Arizona: Westernlore Press, 1996.

Martínez, Oscar J. *Fragments of the Mexican Revolution: Personal Accounts From the Border*. Albuquerque: University of New Mexico Press, 1983.

Mason, Herbert Molloy, Jr. *The Great Pursuit*. New York: Random House, 1970

McGaw, William C. *Southwest Saga*. Phoenix: Golden West Publishers, 1988.

Meed, Douglas V. *Bloody Border*. Tucson, Arizona: Westernlore Press, 1992.

Meyer, Michael C. *Mexican Rebel: Pascual Orozco and the Mexican Revolution 1910-1915*. Lincoln, Nebraska: University of Nebraska Press, 1967.

Miller, Stuart Creighton. *Benevolent Assimilation*. New Haven: Yale University Press, 1982.

O'Connor, Richard. *Black Jack Pershing*. Garden City, New York: Doubleday & Company, Inc., 1961.

Patterson, Richard. *The Train Robbery Era*. Boulder, Colorado: Pruett Publishing Company, 1991.

Pérez, Esther R. and James and Nina Kallas, eds. *Those Years of the Revolution: As Told by Veterans of the War.* San José, California: Aztlan Today, 1974.

Peterson, Jessie and T.C. Knoles. *Pancho Villa: Intimate Recollections by People Who Knew Him.* New York: Hastings House, 1977.

Reed, John. *Insurgent Mexico.* Middlesex, England: Penguin Books, Ltd., 1983

Ruiz, Ramón Eduardo. *Triumphs & Tragedy: A History of the Mexican People.* New York and London: W.W. Norton & Company, 1992

Salamanca, Bonifacio S. *The Filipino Reaction to American Rule 1901-1913.* New Haven: The Shoe String Press of Yale University Press, 1968.

Sawicki, James A. *Infantry Regiments of the U.S. Army.* Wyvern Publications. 1981.

Schirmer, Daniel B. and Stephen R. Shalom. *The Philippines Reader: A History of Colonialism, Neocolonialism, Dictatorship and Resistance.* Boston: South End Press, 1987.

Scott, Major General Hugh Lenox. *Some Memories of a Soldier.* New York and London: The Century Company, 1928.

Sexton, William Thaddeus. *Soldiers in the Sun: An Adventure in Imperialism.* Freeport, New York: Books for Libraries Press, 1971.

Serrano, T.F. and C.O. del Vando. *Victorias Del General Villa.* El Paso, Texas, 1913.

Snyder, Louis L. and Richard B. Morris, Eds. *A Treasury of Great Reporting.* New York: Simon & Schuster, 1949.

Thord-Gray, I. *Gringo Rebel.* Miami: University Of Miami Press, 1960.

Thrapp, Dan L. *Encyclopedia of Frontier Biography.* Glendale, California: Arthur H. Clark Co., 1988.

Tompkins, Colonel Frank. *Chasing Villa.* Harrisburg, Pennsylvania: Military Service Publishing Company, 1934.

Torres, Elías L. *Twenty Episodes in the Life of Pancho Villa.* Austin: Encino Press, 1973.

_____. *La Cabeza de Villa.* Mexico City, 1938.

Toulmin, Colonel H.A. *With Pershing in Mexico.* Harrisburg: Military Science Publishing Co., 1935.

Turner, John Kenneth. *Barbarous Mexico.* Austin: University of Texas Press, 1984 edition.

Vandiver, Frank E. *Black Jack: The Life and Times of John J. Pershing.* College Station, Texas: Texas A&M University Press, 1977.

Walker, Dale L. *Mavericks: Ten Uncorralled Westerners.* Phoenix: Golden West Publishing, 1989.

Wellman, Paul I. *A Dynasty of Western Outlaws.* New York: Bonanza Books, 1961.

Williams, Ben. *Let the Tail Go With the Hide.* El Paso, Texas: Mangan Books, 1984.

Wocester, Dean C. *The Philippines Past and Present.* New York: The McMillan Company, 1930.

Womack, John, Jr. *Zapata and the Mexican Revolution.* New York: Vintage Books, 1970 edition.

Bibliography

JOURNALS

Bacevich, Major Andrew J., Jr. "Disagreeable Work: Pacifying the Moros, 1903-1906." *Military Review*. Vol. LXII. June, 1982.

Braddy, Haldeen. "The Head of Pancho Villa." *Western Folklore*, Vol. XIX, January 1960, No. 1.

_____. "Myths of Pershing's Mexican Campaign." *Southern Folklore Quarterly*. Vol. XXVII, Sept. 1963.

Gerlach, Allen. "Conditions Along the Border, 1915: The Plan de San Diego." *New Mexico Historical Review*. July, 1968.

Welch, Richard E. J. "American Atrocities in the Philippines: The Indictment and the Response." *Pacific Historical Review*. No. 43. May, 1974.

MAGAZINE ARTICLES

Brittain, Don L. "A Civilian with Pershing in New Mexico." *Password Quarterly* Vol. XIV, No. 2, Summer 1969.

Hager, William M. "The Plan of San Diego: Unrest on the Texas Border in 1915." *Arizona and the West*. Winter 1963.

London, Jack. *Colliers Weekly Magazine*. May 5, 1906.

Mason, Gregory. "The Mexican Man of the Hour." *Outlook Magazine*. June 6, 1914.

Meed, Douglas V. "Lieutenant Patton's Raid." *True West Magazine*. May 1997.

_____. "Suicide Charge at Carrizal." *True West Magazine*. Vol. 40, No. 9., Sept. 1993.

_____. "Who Stole Pancho Villa's Head?" *True West Magazine*. August 1996.

Richardson, Tracy. "A Soldier of Fortune's Story." *Liberty Magazine*. Oct. 10, 1925 and Oct. 17, 1931.

Sandos, James A. "The Plan de San Diego: War and Diplomacy on the Texas Border 1915-1916." *Arizona and the West*. Spring, 1972.

"The Cabeza de Villa." *New Yorker*. Nov. 27, 1989.

Index

4th Cavalry 6
6th Engineer Regiment 160, 166
6th U.S. Field Artillery, Battery G 4
13th Calvary Regiment 127, 137, 144
14th U.S. Infantry 6
18th U.S. Infantry Regiment 4
20th U.S. Infantry 12, 13, 14, 20, 23, 24, 25, 26, 30, 107
51st Volunteer Iowa Infantry 1, 3, 4, 5, 6, 8, 9, 10
1911 Federal Neutrality Laws 118

A

Abd-El-Krim 190
Acaponeta River, Mexico 52
Acaponeta, Mexico 57
Acosta, Encarnación 55
Acosta, Simone 94
Agua Prieta, Mexico 111, 118, 126, 127, 156
Aguilar, Lorenzo 90, 91, 92
Aguinaldo, Emilio 2, 3, 13, 15, 32
Akron, The 193
Algeria 195
Allison, Dave 143, 144
American Expeditionary Force 40, 160, 197
American Medical Association 185
American Smelting and Refining Company 182
Amparan, Juan 179
Anenecuilco, Mexico 73
Angeles, Felipe 100
Arce, Francisco Zamora 55
Aristagoras 180
Army News, The 25, 26, 31, 33
Aubert, Trucy 94, 95
Austria-Hungary 152, 159

B

Bagbag River, Philippines 9
Baja California, Mexico 54, 68, 195
Baker, Newton D. 129
Baltimore, U.S.S. 3
banana men 35, 38, 40
Banana Wars, The 35, 38, 48, 87
Barbary Coast 30, 32
barongs 21
Barraza, Jesús Salas 180
Barrymore, John 30
Beaudit, T.J. 68, 69
Bell, J. Franklin 14, 137
Benevolent Assimilation Policy 2, 16
Bisalin, Philippines 20
Blanquet, Aureliano 90, 91, 92, 96
Bluefields, Nicaragua 36
Boers 37, 61, 62, 63
Bois de Taillaux, France 164
bolos 14, 22, 23, 25
Bolsheviks 162
Bonilla, Manuel 41
Boxer Rebellion 12
Boyd, Charles T. 150
Boyle, W.C. 170
Brennan, M. 119
Brophy, Frank 185, 186
Brown, Bryan 182
Brownwood, Texas 191
Bueno Noche mine 56
Buffalo Soldiers 129
Buli, Luis 67
Bush, George H.W. 186
Bush, Prescott 186
Bustillos, Mexico 100

C

Cadillac Automobile Company 68
Calles, Plutarco Elías 184
Calumpit, Philippines 6, 7, 8
Camp Leach, Washington D.C. 166, 169
Campa, Emilio 90, 91
Cananea, Mexico 47

Index

Cannon, Lee Roy 41
Cañon de Mal Paso, Mexico 83
Cape Horn 42
Cárdenas, Julio 141, 142, 145, 146, 147
Carey, Sandeman 164
Carranza, Venustiano 96, 100, 110, 111, 112, 115, 116, 117, 118, 120, 121, 122, 123, 124, 125, 126, 127, 138, 139, 149, 150, 152, 156
Carrizal, Mexico 150
Caruso, Enrico 30
Casas Grandes, Mexico 58, 63, 100
Cavite, Philippines 9
century plant 74
Cesneros, Jesús 93
Chaffee, A.R. 15
Chamberino, New Mexico 62
Chateau Thierry, France 165
Chaulnes, France 164
Checotah, Oklahoma 172
Chiapas, Mexico 181
Chickasha, Oklahoma 171
Chihuahua, Mexico 53, 58, 62, 63, 83, 84, 85, 92, 93, 96, 97, 99, 100, 102, 103, 110, 111, 112, 115, 126, 128, 133, 135, 139, 147, 151, 156, 157, 180, 189, 194
China 10, 11, 12, 13, 196
Chinese Empire Reform Assn 11
Christmas, Lee 35, 38, 39, 40, 171, 196
Churchill, Winston 162
Coahuila, Mexico 53, 92, 96
Cohan, George M. 155
Colonia Dublán, Mexico 133, 148
Colorados 84, 85, 86, 104, 115, 116, 181(See also Red Flaggers)
Columbus, New Mexico 110, 116, 117, 118, 123, 127, 128, 133, 148, 150, 152, 156, 185
Columbus, Ohio 172
Contreras, Fortino 194
Cora Indians 57
Corral de Piedra, Mexico 173
Corral, Alberto 177, 182

Corral, Luz 102, 177
cottas 24, 25
Coyote, Mexico 140, 141
Cuernavaca, Mexico 76, 80
Culberson's Ranch, New Mexico 133
Culiacán, Mexico 47
Curry, Charles F. 31

D

Davis, Richard Harding 33, 36, 37, 41, 108
Davis, W.E. 170
de Valencia, Amparo F. 121
Death Legion, The 73
Decena Trágica 95
Deming, New Mexico 152
Dewey, George 6
Díaz, Porfirio 42, 45, 46, 47, 48, 49, 51, 52, 53, 55, 56, 57, 58, 64, 65, 67, 68, 73, 83, 89, 91, 93, 95, 97, 108, 116, 124, 133
Dobie, J. Frank 99
Doingt, Belgium 164
Dorados 101, 138, 141, 145
Douglas, Arizona 94, 111, 117, 118, 126, 156
Dreben, Sam 37, 38, 40, 87, 88, 90, 92, 109, 138, 160, 181, 195
Drew, John 31
Dunn, H.H. 74
Durango, Francisco R. 184
Durango, Mexico 92, 99, 122, 138, 151, 172, 174, 177, 178
Dutton, Arthur H. 35

E

El Paso Times 42
El Paso, Texas 58, 93, 94, 96, 104, 105, 106, 107, 108, 109, 110, 115, 116, 117, 118, 120, 121, 123, 126, 129, 137, 146, 148, 149, 150, 156, 159, 169, 181, 183, 185, 186, 190, 191, 194
El Reno, Oklahoma 171

El Salvador 37
Elser, Frank B. 146
English, Edmund F. 11, 12
Ernest, Mrs. Gene 185
Espinosa, Martín 56, 57

F

Fall, Albert 124
Fierro, Rudolfo 102, 109, 110, 122, 123
Finch, James A. 161
First Aero Squadron 133, 134, 135
Fort Bliss, Texas 123, 129, 137
Fort Dodge, Iowa xi
Fort McKinley, Philippines 20
Foster, Ramona L. 195
Foulois, Benjamin Delahauf 134
Fountain, Thomas 86, 89, 94
Franklin Mountains 186
Funston, Frederick 32, 108, 149

G

Gallegos, Francisco 83
Galveston, Texas 116, 119, 171
Garibaldi, Giuseppi 63, 108
Garza, Juan 146
Gatling Gun 8, 39, 89
Gentlemen's Agreement 76
Germany 76, 152, 159, 162
Geronimo 6, 186
González, Abraham 99, 100
Goodell, Joe 159
Gordon, Victor 37
Grant, Frederick D. 25, 26
Griensen, Elsa 139
Groce, Leonard 41
Guatemala 37
Guaymas, Mexico 64, 67
Guerrero, Mexico 100
Guerreros 74
Guzmán, Martín Luis 101

H

hacendados 46, 64, 72, 73
Hacienda Refugio, Mexico 91
Haig, Sir Douglas 162, 165
Hall, James Whitney 184
Hall, Lee 123
Harris, Larry A. 186
Hassan, Paglima 24
Hay, John 2
Heath, Frank 116
Henry, O. 171
Hermosillo, Mexico 95, 97, 111
Hernandez, J.H. 120
Herodotus 180
Hill, Benjamin J. 97, 112, 115, 116, 119, 156, 157
Hindenburg, Paul von 164
Histiaeus 180
Hoar, George Frisbie 25
Holmdahl, Gordon xiii, 80, 166, 191, 193, 196
Holmdahl, Monty 1
Honduras 36, 37, 38, 41, 171
Hotel Monteleone 36
Houston, Temple 171
Huerta, Adolfo de la 151
Huerta, Victoriano 78, 79, 86, 89, 90, 92, 94, 95, 96, 97, 100, 102, 103, 106, 111, 116, 120, 125, 141
Hughes, Fred T. 173

I

Iloilo, Philippines 3, 4
International Workers of the World 47
Ipal, Philippines 26

J

James, Milton 128
Jennings, Al 170, 171, 172, 173
Jennings, Ed 171
Jiménez, Mexico 85, 86
Johnson, "Old Pants" 147
Jolo, Philippines 19, 20, 21, 22, 23, 24, 26, 129
Juárez, Mexico 37, 58, 61, 63, 71, 83, 85, 93, 96, 102, 103, 104, 105, 106, 107, 116, 126, 181, 184
Juramentados 21, 22, 23

Index

K
Katz, Friedrich 187
Kelly, Charles 108
Kipling, Rudyard xi
Kosterlitzky, Emiliano 52, 53, 54, 55, 93
Krag-Jorgenson Rifle 12, 13
kris 21, 25, 26

L
La Cucaracha 86
Lantacas 24
Las Nieves, Mexico 122
Lawton, H.W. 5, 6, 13
Lawton, Oklahoma 173
Lea, Tom 120, 123, 159
Leavenworth Federal Prison 159
Leibson, Art 138
Ley Fuga 14
Lloyd George, David 162
London, Jack 29, 32, 33
Longre, France 166
Lopez, Gilardo 147
Lopez, Isadore 146
Los Angeles, California 42, 46, 49, 68, 166, 170, 172, 173, 185, 189, 193
Ludendorff, Erich von 164
Luzon, Philippines 4, 13, 14, 15, 20, 32

M
MacArthur, Arthur 10, 15
MacArthur, Douglas 10
machine guns 24, 39, 90, 100, 104, 162, 163, 165
Madero, Francisco 47, 52, 54, 55, 56, 57, 58, 60, 63, 64, 67, 68, 69, 73, 78, 83, 84, 86, 88, 89, 90, 91, 93, 94, 95, 96, 98, 99, 100, 106, 108, 116, 124, 181
Madrid, Spain 169
Magón, Ricardo Flores 47, 53, 56, 58
Malan, Jack 63
Maloney, Guy 37, 40
Managua, Nicaragua 40, 41, 42
Manila, Philippines 1, 3, 4, 6, 10, 11, 12, 13, 24, 26
Marne River, France 166
Martinez, Oscar A. 184
Maxely, Thomas S. 122
Maxim, Sir Hiram 24, 39, 89, 90, 106
Mayhbun, Philippines 24
Mazanillo, Mexico 42
Mazatlán, Mexico 48, 52
McCoy, Frank Ross 25
McDermott, John 23
McGaw, Bill 52, 137, 148, 183, 195
McKinley, William 1, 2
McLemore, Jeff 157, 158, 160, 161
McPherson, Aimee Semple 191
Medina, Juan M. 97
Mena, Luis 42
Mesilla Valley, New Mexico 62, 63
Mexicali, Mexico 111
Mexican revolution xii, xiii, 35, 54, 86, 94, 96, 104, 110, 112, 117, 178, 181, 194
Mexico City 53, 64, 67, 68, 69, 71, 73, 76, 77, 78, 86, 95, 107, 112, 125, 151, 179, 184, 187, 192, 194
Miller, Owen W. 166
Mindanao, Philippines 19, 20, 23, 129
Mindoro, Philippines 3
Mixtecas 74
Monterey, California 25, 30, 31, 33
Monterrey, Mexico 89, 90, 92
Moore, Mrs. J.J. 127
Morales, Mexico 71, 81
Morelos, Mexico 54, 69, 72, 73, 74, 76, 111, 112
Morenci, Arizona 123
Moros 19, 20, 21, 22, 23, 25
Muskogee, Oklahoma 171

N
Navajo, Mexico 65
Navarrete, Guillermo Rubio 92
Navarro, Juan 63
Nayarit, Mexico 48

Soldier of Fortune

New Mexico 62, 85, 110, 116, 117, 120, 123, 124, 125, 133, 152, 185
New Orleans 36, 38, 40, 42, 192, 196
Nicaragua 36, 37, 41, 42, 191
Nogales, Arizona 47, 72, 92, 95, 111
North, John Ringling 185
Northwestern Railway 135
Nuevo León, Mexico 92

O

Obregón, Alvaro 96, 187
Ochoa, Victor 115, 117, 118, 121, 156
Ojinaga, Mexico 105
O'Reilly, Edward "Tex" 40, 196
Orozco, Jorge U. 115, 121
Orozco, José 115, 121
Orozco, Pascual 58, 63, 83, 84, 85, 86, 87, 88, 89, 90, 91, 96, 102, 120, 181
Otis, Elwell 6
Otomis 74

P

Pala, Datto 24, 25
Palace Hotel, San Francisco 30, 31
Palas Cotta, Philippines 26
Palawan, Philippines 3, 20
Palomas, Mexico 116
Panay Expeditionary Force 4
Panay, Philippines 3
Parque, Mexico 77
Parral (Hidalgo del), Mexico 86, 88, 138, 149, 173, 177, 179, 180, 181, 182, 184, 187
Pasig River, Philippines 13
Patton, George S. xii, 5, 139, 140, 141, 142, 143, 144, 145, 146, 147, 148, 158, 197
Paxinosa, S.S. 173
Peck, Garret 192, 193
Pedriceña, Mexico 91
Peña, Colonel 78
Peña, María 91
Pennsylvania, U.S.S. 1, 3, 4

Pershing, John J. "Black Jack" xii, 23, 108, 128, 129, 130, 133, 135, 136, 138, 139, 140, 146, 147, 148, 149, 150, 151, 152, 157, 158, 180, 194, 197
Philippine Insurrection 5, 36, 39
Phoenix, Arizona 185
Pimas 68
Plan de Empacadora 84
Plan of Ayala 73
Plan of San Diego 125, 127
Polk, Frank 159, 160
Porfiriato 46
Presidio at Monterrey, The 25, 30, 129
Proctor, Redfield 3
Punitive Expedition, The xii, 133, 139, 146, 148, 150, 158, 181
punjis 8
Punta Banda, Mexico 195

Q

Quingua River, Philippines 6
Quingua, Philippines 7

R

Ramos, Héctor 117, 120
Rancho Guerachic, Mexico 99
Ravel, Sam 127
Read, G.W. 161
Red Flaggers 84, 86, 89, 91, 93, 94, 121 (See also Colorados)
Reed, Thomas B. 2
Regeneración 47, 53
Rellano, Mexico 86, 90
Richardson, Tracy 37, 38, 40, 87, 88, 90, 92, 108, 160, 181, 186, 192
Riff guerrillas 190
Ringling Brothers Circus 185
Rio Grande 93, 96, 100, 105, 107, 108, 111, 120, 125, 186, 187
Robertson, Sam 160
Robles, Juvencio 71, 79
Roosevelt, "Teddy" 10, 20, 129
rope cure 14
Rosamorada, Mexico 51, 56

Index

Rosario, Mexico 56, 173
Rough Riders 20, 129
Rubio, Mexico 139, 140, 141, 142, 146
Rurales 49, 50, 51
Ryan, J.A. 157

S

Sage, Russell 137
Salas, González 86, 87, 89
Salazar, Inez 86, 94, 96, 97, 103, 105, 108, 120
Salsona, Philippines 14
San Andrés, Mexico 100, 102
San Diego, California 68, 195
San Fernando, Philippines 8, 9
San Francisco, California 26, 29, 30, 32, 33, 36, 105, 106, 129, 172
San Juan Hill, Battle of 10, 129
San Juan River, Nicaragua 41
San Mateo, Philippines 13
San Miguelito Ranch, Mexico 141, 147
Santa Rosalía, Mexico 85
Schmitz, E.E. 32
Scott, Hugh L. 23, 24, 108, 110, 123, 124, 125, 135, 152, 159
Scott, Orlando F 184
Seagraves, O.R. 117
Senator, U.S.S. 10
Service, Robert 197
Shadbolt, L.M. 186
Shallenberger, M.C. 148, 149
Shanghai, China 11, 12, 191
Sheldon Hotel, El Paso 107, 108, 109, 110, 115, 159, 185, 186
Sheppard, Morris 161
Sheridan, U.S.S. 26
Sherman, William Tecumseh 9
Sierra Blanca, Texas 144
Sierra Madres 126, 134, 138
Sinaloa, Mexico 47, 58, 65, 68, 76, 97, 121
Skull and Bones Club 185, 186
Slocum, Herbert 137
Smith, Jacob W. "Hellroarin' Jake" 14, 29

Soldado de Fortuna 194
Sonora, Mexico 47, 52, 53, 58, 64, 65, 66, 67, 68, 71, 84, 92, 96, 97, 118, 126, 156, 188
South China Sea 3
Spanish-American War x, 20, 39, 129
St. Cyr, France 100
Stewart, William C. 170, 172, 189, 190
Stotsenberg, John M. 7
Sulu Archipelago 20
Sulu Island 24
Sulu Sea 3, 19, 20, 21

T

Taft, William Howard 15, 42, 124
Talai, Philippines 23
Tamaulipas, Mexico 92
Tambang Market, Philippines 26
Tampico, Mexico 86, 124
Taylor, H.L. 160
Tegucigalpa, Honduras 41
Temple, H.J. 95
Tepic, Mexico 48, 49, 56, 57, 58, 65, 68, 69, 76
Terrazas, Félix 100
Terrazas, Luis 84
Texas 58, 85, 93, 100, 106, 107, 108, 109, 112, 116, 123, 125, 143, 148, 149, 155, 157, 160, 171, 191, 194
Third Sulu Expedition 26
Thord-Gray, I. 65, 103, 104
Tía Juana, [Tijuana] Mexico 111
Tierra Blanca, Mexico 103, 106, 111
Tlahuicas 74
Todos Santos Bay 195
Tompkins, Frank 128, 138, 139
Torreón, Mexico 75, 89, 90, 102
Torres, Elías L. 184
Toulmin, H.A. 130
Treviño, Gerónimo 89, 90, 92
Turner, John Kenneth 65
Tuxpán, Mexico 50
Tzu Hsi 11

U

U. S. Bureau of Investigation 94, 117
United Fruit Company 36
Urbina, Tomás 122, 170, 172, 173, 177

V

Vaalkrantz, Battle of 61
Van Nuys, California 186, 194
Vasey, Peter 23
Vera Cruz, Mexico 86, 119, 120, 124
Vieux Carré 36
Viljoen, Benjamin 61, 62, 63, 64, 67, 68
Villa Ahumada, Mexico 150
Villa, Pancho xi, 58, 84, 85, 96, 99, 110, 111, 123, 128, 152, 177, 178, 179, 184, 186
Visayas, Philippines 3

W

Walker, William 145
Warren, Francis E. 129
Warren, Helen Frances 129
Washington Barracks 160
water cure 14
Waterfill, Mexico 194
Wells Fargo 171
West Point 10, 129, 146, 150
Western Front, The 162, 164
Wheeler, Joseph 9
Wilkins, Sir Hubert 193
Williams, Ben F. 185
Wilson, Henry Lane 95
Wilson, Woodrow xii, 124, 160, 196
Wood, Leonard 20, 24, 25
Wylie, Elinor 169

Y

Yale University 185
Yaquis 60, 61, 64, 65, 66, 67, 68, 69, 73, 97, 99, 185, 196
Ysleta, Texas 106
Yucatán, Mexico 66, 67, 68

Z

Zacatecas, Mexico 92
Zamboanga, Philippines 20
Zapata, Emiliano 69, 71, 72, 73, 74, 75, 78, 79, 80, 81, 92, 95, 111, 112
Zapatistas 73, 74, 75, 76, 77, 78, 79, 80, 121
Zapotecas 74
Zaragoza, Mexico 107
Zelaya, José Santos 41, 42
Zemurray, "Sam the Banana Man" 36